Analysis for Residuals-Environmental Quality Management

According to the laws of conservation of mass and energy, matter cannot be created or destroyed. Therefore, all human activities result in some residuals. In this title, originally published in 1978, the authors describe the overall magnitude of the annual residuals problem and apply a residuals-environmental quality management (REQM) analysis specifically to the industrial development of the Ljubljana area in the former Yugoslavia. This title is ideal for students interested in environmental studies and international development issues.

Analysis for Residuals-Environmental Quality Management

A Case Study of the Ljubljana Area of Yugoslavia

Daniel J. Basta, James L. Lounsbury, and Blair T. Bower

RFF PRESS
RESOURCES FOR THE FUTURE

First published in 1978
by Resources for the Future, Inc.

This edition first published in 2016 by Routledge
2 Park Square, Milton Park, Abingdon, Oxon, OX14 4RN
and by Routledge
711 Third Avenue, New York, NY 10017

Routledge is an imprint of the Taylor & Francis Group, an informa business

© 1978, Resources for the Future, Inc.

Publisher's Note
The publisher has gone to great lengths to ensure the quality of this reprint but points out that some imperfections in the original copies may be apparent.

Disclaimer
The publisher has made every effort to trace copyright holders and welcomes correspondence from those they have been unable to contact.

A Library of Congress record exists under LC control number: 77017250

ISBN 13: 978-1-138-95810-4 (hbk)
ISBN 13: 978-1-315-66129-2 (ebk)
ISBN 13: 978-1-138-95823-4 (pbk)

Analysis for residuals-environmental quality management

A case study of the Ljubljana area of Yugoslavia

DANIEL J. BASTA, JAMES L. LOUNSBURY,
and BLAIR T. BOWER

RESEARCH PAPER R-11

RESOURCES FOR THE FUTURE / WASHINGTON, D.C.

Resources for the Future is a nonprofit organization for research and education in the development, conservation, and use of natural resources and the improvement of the quality of the environment. It was established in 1952 with the cooperation of the Ford Foundation. Grants for research are accepted from government and private sources only if they meet the conditions of a policy established by the Board of Directors of Resources for the Future. The policy states that RFF shall be solely responsible for the conduct of the research and free to make the research results available to the public. Part of the work of Resources for the Future is carried out by its resident staff; part is supported by grants to universities and other nonprofit organizations. Unless otherwise stated, interpretations and conclusions in RFF publications are those of the authors; the organization takes responsibility for the selection of significant subjects for study, the competence of the researchers, and their freedom of inquiry.

Library of Congress Catalog Card Number 77-17250
ISBN 0-8018-2088-X

Copyright © 1978 by Resources for the Future, Inc.

Manufactured in the United States of America

Published June 1978.

FOREWORD

Two simple things distinguish this study. Contrary to what is commonly
expected, they are neither the numbers nor the modern jargon of environmental
analysis. Rather they are: first, a demonstration that a simple framework
of analysis can be adapted to very different conditions from those in which it
was originally created; and second, a cross cultural context, rather than
being a barrier to learning, can provide a demonstration of the value of such
learning. Drawing on the truth that to live is to pollute, this study suggests
a framework within which decision makers in widely different kinds of societies
can approach the common problem of "environmental quality management".

No society is free of pollution. Because matter and energy are neither
created nor destroyed, but merely transformed by human activities, residuals
which society chooses to discharge rather than to reuse, at any given time,
must be returned to the environment. There is no other place for them to go.
These residuals may be liquids, solids, gases, heat or other forms of energy.
In virtually all societies today where large numbers of people and industrial
activities are concentrated in cities and towns, pollution from these activities
is concentrated in a limited region. Thoughtful analysis leading to the choice
of appropriate management strategies is essential if the system involving
society's activities, including the disposal of its residuals to the environment, is to be managed in such a way that the environment itself can continue
to satisfy the needs and desires of people within the region.

Only in the very recent past have citizens, politicians, engineers, and scientists begun to grasp the meaning of these seemingly obvious truths. Over much of the globe in the historical past, it has been possible to act as if wastes could be handled by the old adage, "out of sight, out of mind". The middens of antiquity suggest that the objective of the adage was probably never realized. But even if it had been, it is clearly not possible in most parts of the world today to ignore the problem of wastes. There are too many of us, we choose to live together, and we draw upon vast areas for food, clothing, shelter, and the products of our labor.

Recognition that an urban region cannot escape from the necessity of dealing with the potential problems of environmental pollution evokes a demand by government for a framework within which government, business, and individuals can assess both the magnitude of potential environmental quality problems and alternative strategies to assure the achievement and maintenance of satisfactory environmental quality. The residuals environmental quality management study described herein provides a framework for such analysis. It includes analysis of the flows of materials and energy in the many activities distributed over a region and evaluation of the environmental quality impacts and of the opportunities and costs of alternative strategies which might be employed to achieve different levels of environmental quality.

Basta, Lounsbury, and Bower and their colleagues in Ljubljana have shown that the essential task of analysis, a prerequisite to intelligent choice of environmental management policies, can and must be adapted to the special conditions of the local scene. They show that the data needed, the models to be used, and the alternative strategies to be considered are functions of what data can be best obtained, what methods of analysis will be sufficient, and what options are best suited to the needs of real decision makers. Decisions must be made not only about the management strategies themselves, but also about the kinds of data needed and the sophistication of analyses to be used. All of these decisions must be set by the possibilities for action and by the limits of budgets. Different societies will not only have different environmental quality objectives but they will differ in the resources, including men and women, skills, and money, which they can apply to their problems.

Because environmental quality management is required not only of the most developed and industrialized nations but of the entire spectrum of nations, we are indebted to the authors of this report for illuminating both the opportunities available to decision makers and the choices they can exercise in matching demands to problems, pocketbook, and available skills. The study recognizes that different societies do things differently, but it also recognizes that all human beings and societies in striving to fulfill their needs must seek ways to maintain and enhance the environment within which they live. The questions which need asking are universal; the answers particular. The Ljubljana study suggests a framework for asking questions, and illustrates some choices and ways of proceeding to answer them. Hopefully it will stimulate others in a wide variety of settings to ask the basic questions, to set about ways of seeking answers, and to make those choices which will lead to the healthy and aesthetic environment so universally sought.

M. Gordon Wolman

PREFACE

Environmental pollution problems are endemic to all countries, whatever the stage of economic development. Increasing attention has been given to such problems over the last decade. In many cases the focus of attention has been on some region--a metropolitan area, a river basin, a coastal area--where the processes of urbanization-industrialization-tourism have led to overexploitation of the environment and subsequent deterioration of ambient environmental quality. Such situations occur in market and non-market, so-called unplanned and planned economies alike. Whatever the context, decisions must be made on how to cope with environmental degradation consistent with the particular political-economic system in which those decisions are made, and with the social goals of that system.

Decision making takes place with more or less rigorous analysis, perhaps in some cases with essentially no analysis at all. The study reported herein was undertaken with the beliefs that first, better decisions concerning management of a society's residuals can be made if more information on the costs and consequences of alternative management strategies is available, and second, analytical methods for generating such information exist and can be applied.

Undertaking analysis to provide information on the costs and consequences of alternative strategies for improving ambient environmental quality is a difficult task in any context, but is even more complicated in a cross-cultural context. The study for residuals-environmental quality

v

management in the Ljubljana area was done by an analysis team comprised of individuals from two different cultures, Yugoslav and American.

That the study was undertaken in Ljubljana, Yugoslavia was a result of the conflation of two factors. The first was the existence during the period of 1966-76 of a joint U.S.-Yugoslav research program on urban problems at the Urbanisticni Institut of the Socialist Republic of Slovenia in Ljubljana. The U.S. institution involved was the Center for Metropolitan Research and Planning of Johns Hopkins University. This existing program provided the institutional base for the study. The second was the existence of a program of research on regional residuals-environmental quality management at Resources for the Future, Washington, D.C., during the same period. This provided the analytical base for the study. The study was funded primarily by grants of U.S.P.L. 480 funds to the Center for Metropolitan Planning and Research, supplemented by a small grant from Resources for the Future. Support for preparation of the final report was provided by Resources for the Future.

The project was organized to be as Yugoslav oriented as possible, in order that the results would in fact be useful. This meant attempting to maximize: the participation of Yugoslav staff members; the development and utilization of data specifically reflecting Yugoslav conditions; and the development and analysis of management strategies which were perceived by the Yugoslavs as being relevant to their conditions. Such a cross-cultural, truly joint effort is no easy task; this report reflects some of the problems associated with that type of endeavor.

Critical Factors in Cross-Cultural Studies

Reflecting on this cross-cultural experience leads logically to the attempt to identify factors which are particularly important if a successful study is to be achieved in a cross-cultural context. Some of these factors are important in any context, but their importance is increased in a cross-cultural setting. Just as the Ljubljana study was not the first nor the last cross-cultural effort, so the following list of factors is neither the first nor the last effort to provide some guidelines. The list represents our attempt, at the behest of M.G. "Reds" Wolman, to appraise our experience.

1. Defining the scope of work to be performed is a step which often receives inadequate attention. The task must be a joint enterprise of both groups. It must clearly identify the level of problem solving involved, the study area boundaries, the goals of the study, the specific questions to be answered, and the specific nature of the end product or products to be produced for whom in the time alloted. It is important to emphasize to all individuals involved that several iterations of this phase are likely to be necessary, as a more clear conception of level of problem solving, goals, relevant study area, and outputs evolves in the early period of the analysis.

2. One or more agencies which will use the outputs of the study must be involved. There will be little utility in performing a study, other than as an academic exercise, without there being some link to the decision-making process. At least some of the individuals involved from the host country should be those who will be responsible for, or involved in, continuing the effort after the cross-cultural phase is completed.

3. The work effort itself must be a joint undertaking, with both sides being guaranteed mutual acceptance in the working relationship. Full-time qualified personnel who are competent to perform the tasks specified under the scope of work, must be supplied by both sides. It is particularly important that full-time qualified personnel from the host country be available. An imbalance in the numbers or qualifications of the staff can seriously affect the rate and quality of work progress and the credibility of the output to the local users.

4. In addition to its being essential that there be a joint, cross-cultural working team, host country personnel have two other essential roles. One is to insure that the assumptions and constraints relevant to the particular cultural context in which the study is being undertaken are set forth accurately and explicitly. For example there may be: governmental policies which limit the use of foreign technology; constraints on the availability of foreign exchange to purchase equipment; a particular procedure for estimating cost of land; special considerations which determine the time value of money to be used in the estimation of costs; cultural patterns which affect final demand and the choice of implementation incentives and institutional arrangements. Often it is the "foreigners" who stimulate the process and push for clarification, but only the host country professionals can explain the nuances. The utility and credibility of a study can be seriously diminished if these considerations are not set forth explicitly by the joint team.

The other role is to suggest, locate, and exploit sources of primary data. Locations where data are stored and procedures for unlocking the doors thereto are peculiar to each culture. The often mysterious ways in which data can be obtained from any bureaucracy in a country are usually known only by host country personnel, and then often by only a few imaginative individuals. Coopting individuals and agencies to obtain primary data requires particular finese in a cross-cultural context.

5. Control of the resources for the study must be in the hands of those actually doing the work "on the ground". This is because shifts in resources among study components will inevitably be required as the study proceeds. It often is not clear at the start of a study, particularly in a cross-cultural context, what empirical data are actually available and/or if there will be any constraints on obtaining and using certain data. As in any study, perceptions and understanding of the problems often change in the course of a study. Failure to structure a project administratively so as to provide for flexibility in allocating project resources will inhibit achieving the goals of the project.

6. Doing a study in a rapidly developing economy places a premium on explicit consideration of the dynamic nature of the context and of the multiplicity of uncertainties involved. Because technology, factor prices, social tastes, and governmental policies have major impacts on management strategies and on costs, it is important for the analyst to look at alternative assumptions about such variables, if the information generated in the study is to convey adequately to the decision makers the implications of the uncertainties associated with those variables.

The multiplicity of uncertainties makes a first round-second round
analysis procedure essential, within the available time for a study. This
is particularly true if the type of study is being done for the first time.
Goals, objectives, characteristics of outputs are all likely to require
modification after the first round.

7. A perennial problem is producing an end product in the time
allotted, particularly one which is useful to the host country. Determining
what the nature of that product should be is itself a major and critical
task. Too often inadequate consideration is given to what data and results
are to be presented, in what form, to whom, addressing what questions.

This list is by no means exhaustive, but represents problems which
beset us.

Nature of the Report

The report carries the reader through the complete process of analysis,
from the first structuring of the residuals-environmental quality management problem in the Ljubljana area, through the empirical data collection
consisting of sampling liquid and solid residuals generated by various
activities in Ljubljana and the estimation of the costs of various physical
measures for reducing residuals discharges, to the delineation and analysis
of combinations of physical measures in relation to environmental quality
targets.

As the reader travels the rocky road of this discussion he will note
the many decisions which had to be made, given limited analytical resources,

limited data, and a finite time period for the study. Hopefully the description of these decisions will provide some insight into the "real world" tradeoffs which affect the analyst: tradeoffs between data collection and the amount of information obtained, between data collection and costs of data collection, between data collection and accuracy of outputs, between putting more resources into the analysis of one component and fewer resources into the analysis of another.

Two additional points concerning the report merit emphasis. One, the report demonstrates what can be done to produce useful information for decisions with limited analytical resources, few existing primary data, and relatively unsophisticated analytical procedures. Little utility can be gained by applying sophisticated analytical methods where limited data are available and where no institutional structure exists which is accustomed to using and has the authority to use the outputs of such methods. Two, the report reflects the fact that the approach to the analysis of REQM presented herein is apolitical. Whatever the structure of a society, levels of AEQ desired and strategies for achieving them--or trying to achieve them-- will be chosen. The procedure for generating the needed information is generic; the goals, social values, constraints, analytical resources differ from society to society.

Finally, it should be recognized that analysis is only one part of the total management process of producing improved ambient environmental quality. Analysis by itself never improved any dimension of ambient environmental

quality. But analysis is the beginning. It is hoped that this study is a
small beginning toward development of a continuing capacity in the Ljubljana
area to generate information for management decisions, which can be implemented by Yugoslavs to produce improved ambient environmental quality in
their country.

D.J. Basta

B.T. Bower

ACKNOWLEDGMENTS

It would have been impossible to have done this study without the cooperation of, and inputs from, many individuals and agencies. First and foremost are the Urbanisticni Institut of the Socialist Republic of Slovenia, the host institution in Ljubljana, B. Music, Director, and all its members, and the Johns Hopkins University Center for Metropolitan Planning and Research, J. Fisher, Director. Very special thanks are due Joze Dekleva of the Urbanisticni Institut, whose tireless efforts and persistence were indispensable in bringing the study to completion. Had the timing of and the physical proximity for, preparation of the final report been different, Dekleva would have been a co-author of this report. Special thanks are also due: Franc Vrecl of the Urbanisticni Institut for his work on the study; Walter O. Spofford, Jr. and William J. Vaughan for their thorough, detailed review of the manuscript; M. Gordon Wolman for his continued encouragement, many cogent comments, and suggestions throughout the effort; Charles Ehler for his review of the manuscript; Kerry Smith for his review of sections of the manuscript; Luz Maria Aveleyra for her patience and skill in typing the final manuscript; Pathana Thananart for ably drafting the charts and figures; and Margaret White and John Mankin for typing earlier drafts of the manuscript. Listed below are the project staff and the various individuals and agencies who contributed to the effort.

Project Staff (1973-1975)

Project Director: D. Basta

Program Manager: J. Lounsbury

Research Associates: J. Dekleva
                      F. Vrecl
                      L. Koss

Research Assistants: A. Deticek
                      S. Djurasevic
                      S. Jakin
                      Z. Mravljak
                      M. Sagadin
                      E. Sefer
                      B. Sverko
                      M. Zitko

Computer Programmer: D. Cepar, Institut Jozef Stefan

Secretarial: G. Kazazic, Urbanisticni Institut SRS, Ljubljana
                      D. Sullivan, Johns Hopkins University, Baltimore

Consultants: Dr. T. Angotti, Rome
                      Mr. B. Bower, Resources for the Future, Washington, D.C.
                      Dr. H. Day, College of Environmental Sciences,
                        University of Wisconsin-Green Bay, Green Bay
                      Dr. T. Lakshmanan, Department of Geography and En-
                        vironmental Engineering, Johns Hopkins University,
                        Baltimore

Individuals in
Ljubljana, Yugoslavia: Dr. B. Music, Director, Urbanisticni Institut SRS
                      Dr. J. Kolar, Komunalno Podjetje Kanalizacija
                      J. Gosar, Komunalno Podjetje Kanalizacija
                      Dr. J. Petkovsek, University of Ljubljana
                      Dr. B. Paradix, Meteroloski Zavod SRS
                      Dr. H. Hocevar, University of Ljubljana
                      Dr. J. Gruden, University of Ljubljana
                      Dr. P. Novak, University of Ljubljana
                      Dr. Pavletic, University of Ljubljana
                      M. Ivanc, SMELT Consulting
                      P. Solar, Komunalno Podjetje za Energetiko KEL
                      D. Botina, Komunalno Podjetje SNAGA
                      H. Senekovic, Zavod za Vodno Gospodarstvo

Individuals in the
United States:

Dr. J. Cohon, Department of Geography and Environmental Engineering, Johns Hopkins University,
Baltimore

Dr. J. Fisher, Director, Center for Metropolitan
Planning and Research, Johns Hopkins University,
Baltimore

Dr. W. Spofford, Director, Quality of the Environment
Division, Resources for the Future, Washington, D.C.

Dr. M. Wolman, Chairman, Department of Geography
and Environmental Engineering, Johns Hopkins
University, Baltimore

Agencies and Firms in
Yugoslavia:

Mestni Vodovod
Splosna Vodna Skupnost
Surovina
Snaga-Kommunalno Podjetje Ljubljana
Dinos
Hidrometeoroloski Zavod SRS
3melt
Kanalizacija-Kommunalno Podjetje Ljubljana
Zavod za Vodno Gospodarstvo
Cestno Podjetje
Ljubljanska Banka
Ruthner-Wien (Vienna, Austria Office)
Institut Jozef Stefan
Pivovarna Union
Sekretariat za Gospodarstvo SRS, Inspekcija za Kotle
Toplarna Moste
Toplarna Siska
Petrol
Kurivoprodaja
Stanovanjsko Podjetje FOND
Stanovanjsko Podjetje DOM
Istra Benz
Kurivo
Zavod za Statistiko SRS
Institut za Zdravstueno Hidrotehniko
Zavod za Regionalno Prostorski Planiranje
Kemijski Institut Boris Kidric
Fakulteta za Strojnistvo, University of Ljubljana
FNT-Oddelek za Meterologijo
Viator

TABLE OF CONTENTS

List of Tables

List of Figures

Chapter I

INTRODUCTION

The Why of the Study

Why are there problems of environmental pollution? Basically the
reason is as follows. Traditionally in all economies it has been considered that the production of goods and services requires some combination of labor, raw materials including land, and capital. However, one
factor input essential for such production is missing in this formulation, namely, the environmental services used, both as inputs--such as
air for combustion processes--and as depositories for residuals.[1] Because no production process is 100 percent efficient in transforming raw
material inputs into desired outputs of products and services, the disposal services of the environment are absolutely essential. The same
types of environmental services are required by human beings to enable
their personal activities. Thus, all human activities result in the
generation of some residuals.

As long as population was small and dispersed, the assimilative
capacity of the environment was sufficient to take care of the residuals
discharged to the environment without impairment of ambient environmental
quality, or at least without adverse impacts on other activities, even

---

[1]Succinctly, residuals are unwanted materials and energy. Chapter
II contains a more detailed definition and discussion.

though ambient environmental quality (AEQ)[2] might have decreased. When human activities became sufficiently concentrated, the discharge of residuals from one activity began to have adverse effects on the uses of the environment by other activities.

Because the environmental services inputs to production processes and other economic activities were not recognized, and hence had no price attached to them, that combination of labor, raw materials, capital, and environmental services was used which did not constrain the use of what was considered to be the least expensive factor, environmental services. Consequently, more of that factor was used than would have been used if the environmental services had been priced to reflect the social costs of their use, namely, the adverse impacts on other users of the environment. This is the root of past and present environmental pollution problems.

A related misconception has contributed to the lack of understanding of residuals problems. Traditionally, again in all societies, the terms "consumer good" and "consumption" have been used, implying disappearance. In reality, goods only provide service or utility or satisfaction for a shorter or longer period of time. Sooner or later a car, an appliance, a suit of clothes, even a building, no longer provides the

---

[2]The phrase "ambient environmental quality" is used throughout to refer to the environment outside of, or external to, the dwelling unit, work place, work site, in contrast to the environment inside the dwelling unit, work place, work site. Problems of the "inside" environment have generally been the domain of industrial hygiene or occupational health.

desired service, and it is discarded or "thrown away." But there is no "away"; its mass does not disappear, the same quantity of material which went into the product is still in existence and must be disposed of in some manner. The laws of conservation of mass and energy have not been overthrown in any society. This fundamental fact indicates the magnitude of the annual residuals problem: the total weight of residuals to be handled is equal to the total weight of material inputs to production and so-called consumption activities plus the weight of air withdrawn from the atmosphere for use minus the net accumulation of materials in capital goods, such as buildings.

These "facts of life" indicate that dealing with residuals is a management problem, one involving human activities as residuals generators and dischargers, the natural environment, and institutions. Coping with the problem requires a framework both for analysis and planning and for the total task of management, of which analysis/planning is a part. A framework should enable answering such basic questions as: What residuals are being generated and what are their effects on AEQ when they are discharged? What are the damages and/or benefits from changes in AEQ? What activities at what locations generate these residuals, in what quantities, in what time patterns? How can the discharges of these residuals to the environment be reduced to lessen negative impacts? How much will reducing the discharges cost, and who will pay those costs? Who benefits from the resulting changes in AEQ?

The study reported herein describes the application of one analytical framework for answering these questions, namely, residuals-environmental quality management (REQM). (That framework is defined in the next chapter.) To date the REQM framework has been applied mainly to areas in "developed" countries, such as the Lower Delaware Valley of the United States (U.S.)[3], where substantial amounts of data existed and substantial financial and analytical resources enabled use of sophisticated computer modeling techniques. Successful application in such areas still left the question, to what extent can a comprehensive REQM analysis be made operational and what relevant questions for REQM can be answered in areas where: (1) the economic, political, and social context is different than in the U.S.; (2) resources and data for analysis are limited; and (3) economic and political structures are not clearly defined or are non-existent with respect to environmental quality management? To attempt to answer that question the REQM analysis of the Ljubljana, Yugoslavia, area was initiated in 1973 under the auspices of the Center for Metropolitan Planning and Research of The Johns Hopkins University and Resources for the Future in the U.S., and the Urbanisticni Institut of the Socialist Republic of Slovenia in Yugoslavia.

REQM in the Yugoslavian Context

Yugoslavia is a country where industrial development and the demand

---

[3]See W.O. Spofford, Jr., C.S. Russell, and R.A. Kelly, Environmental Quality Management: An Application to the Lower Delaware Valley, RFF Research Paper R-1 (Washington: Resources for the Future, 1976).

for goods and services have expanded rapidly and have only reached substantial levels in the last two decades. Table 1 shows the levels of
several economic indicators for 1962 and 1972.

Table 1. Yugoslav Economic Indicators for the Years 1962 and 1972

	1962[a]	1972[a]	Ratio 1972/1962
Total Production	332	750	2.3
Industrial Production	103	285	2.8
Personal Consumption	166	413	2.5
Employment	260	280	1.1
Population	112	125	1.1
Real Personal Income	+5.4% per yr. (1952-1962)	+5.5% per yr. (1962-1972)	---
Balance of Trade (Exp./Imp.)	150	768	5.1

[a]1952 = 100

Source: "Yugoslavia's Socio-Economic Development, 1947-1972," Yugoslav
Survey, vol. XV, n. 1 (February 1974).

Between 1962 and 1972, the population and labor force of Yugoslavia
increased only slightly, whereas total production (Gross Domestic Product) more than doubled. Both industrial production and personal consumption increased more rapidly than total production. In addition to the
large increase in internal demand for goods and services, the ratio of
exports to imports increased by a factor of five during the period.

The increase in residuals generation and discharge resulting from
the increased level of economic activity did not produce noticeable and
perceived adverse impacts on AEQ until the mid 1960's. But the emphasis

on economic growth and increased output of goods has, for the most part,
not included an associated program for the management of the resulting
residuals. Legal provisions concerning AEQ have been seriously lacking
at all levels of government in Yugoslavia. However, over the last decade,
as citizens and planners have become increasingly aware of deteriorating
AEQ and the effects thereof, some significant institutional changes have
occurred.

In contrast to the U.S., where federal, state, and local governments
are involved in environmental legislation, decision making and management
conditions in Yugoslavia are much different. Basic to these differences
is that Yugoslavia is a federation of six republics: Slovenia, Croatia,
Bosnia, Serbia, Montenegro, and Macedonia. The three governmental levels
are federal, republic, and commune. Most governmental responsibilities
are delegated to the republics and particularly to the communes, the latter
being similar in size and function to U.S. counties.

Governmental involvement in environmental issues in Yugoslavia
dates back only to the early 1960's. In Slovenia it was precipitated by
a controversy over a proposed dam across a scenic and culturally significant mountain valley, which resulted in a resolution by the republic
assembly opposing the dam. By 1965, the federal assembly had passed basic
laws[4] dealing with water and air pollution, and other "guideline" laws

---

[4] A basic law at the federal level is one which sets forth general
goals which are to be adopted by each of the republics.

were in effect concerning the quality of the work environment. Yugoslavia's involvement in international preparations for the United Nations
environmental conference at Stockholm in 1972 was enthusiastic, and de-
scriptions of world and local environmental problems were increasingly
provided in the four Yugoslav languages of Slovenian, Serbo-Croatian,
Montenegran, and Macedonian. Local interest groups formed, introducing a
more persistent thrust than any institution had previously. In Slovenia,
the Skupnost za Varstvo Okolja v Slovenija (Group for the Protection of
the Environment in Slovenia) was founded in 1970, with headquarters in the
Republic's capital, Ljubljana. Independent local groups were organized
in the neighboring cities of Maribor and Koper in 1971.

Recognizing the value of this kind of activity, in 1969 the federal
assembly charged its commission on urban and regional planning--now the
Commission for Urban and Regional Planning and Protection of the Environ-
ment--with preparing drafts of federal laws to outline the environmental
policy of the nation. In 1973, a prestigious "Council for the Preservation and Improvement of Man's Environment" was established at the federal
level, with strong political and government support. This Council ad-
vises the federal government, interest groups, and other organizations
throughout Yugoslavia.

To date, the federal assembly has adopted a statement on "urbaniza-
tion, environment, and spatial organization", revised the water pollution
law, and ratified international agreements on the disposal of nuclear
wastes and the protection of birds. At the time of the study (1973-75),

the federal assembly was considering drafts of laws involving herbicides, hazardous materials, pipelines, and additional aspects of the protection of running water and the seas. Much of this activity proceeds directly from Article 87 of the 1974 Federal Constitution and Articles 73-75 of the previous (1963) constitution.[5]

In general, however, the federal government does not implement any environmental legislation; its function is strictly to develop and promulgate general guidelines. Governments at the republic level attempt to implement federal guidelines by establishing commissions and programs encouraging research on and implementation of measures to improve AEQ, but the republics still do not supply significant resources for these functions.

The task of implementing and enforcing AEQ standards and developing and carrying out REQM is delegated to the communes. Under this decentralized system each commune in theory is responsible for maintaining the quality of the local environment, consistent with federal and republic guidelines. Implementation is by a workers-management decision-making framework in which a representative council of all workers' activities in a commune collectively determines, implements, and enforces local policy for the improvement of AEQ. However, there has been little incentive as

---

[5] See T.J. Wilbanks, *Environmental Management with a Really Decentralized Government* (Ljubljana, Yugoslavia: Syracuse University Environmental Policy Project, 1973).

yet to allocate the necessary resources for implementation, given the overriding goal of economic development, particularly industrialization.

This institutional system for implementing environmental policies, where the primary responsibility is at the commune level, does not provide an institutional structure for handling REQM problems which involve more than one commune, although communes can voluntarily form regional agencies to which specific powers can be transferred.[6] In Slovenia there are two operating agencies which can affect REQM on a regional basis. One is SEPO, a division of Ljubljanska Banka, which has adopted a policy of not approving credit for projects with negative environmental impacts. This agency makes its decisions based on limited residuals generation and project design information submitted by prospective clients, and it has won most of the legal contests which have been precipitated over requiring the inclusion of residuals modification measures in project design proposals.

A second is Splosna Vodna Skupnost, which is a technical arm of the Slovenian Republic Water Quality Commission and is funded by the workers-management organizations within the communes of the Upper Sava River basin. To date Skupnost has forced the construction of liquid waste treatment facilities in 60 of the several hundred industrial operations in Slovenia which discharge liquid residuals into the Sava River or its tributaries.

It is likely that it will be several years before Yugoslavia's

---

[6]Ibid, pp. 62-63.

policy on environmental decision making and REQM crystallizes. However, the authority for communal agencies to establish broader regional (multi-commune) agencies appears to provide a viable political vehicle for solving regional REQM problems within the self-management structure that presently exists. The main problem becomes: is the desire to achieve a "cleaner" environment strong enough to motivate that vehicle?

Goals of the Study

Many factors influence goal selection in any given context. Two principal ones influenced the selection in the Ljubljana study. The first was the desire to generate information on costs of improving AEQ to different degrees and by different means. Most societies today are becoming relatively more concerned with environmental quality problems and consequently are seeking ways to develop strategies to prevent further deterioration in and to improve AEQ, in relation to other societal goals. Despite differing political and economic systems, the principal questions are: what are the least cost measures to improve AEQ; how much improvement in what kinds of AEQ can be purchased for what expenditure of resources; who pays; and who benefits? Therefore, the first factor in selecting goals for the study was to shed light on these questions in such a fashion as to provide useful information to the Yugoslavs.

The second factor was the conspicuous lack of a formally constituted user agency, that is, one which could use the generated information

directly in management decisions. In contrast to many studies performed
to respond to some specific set of questions asked by some defined user,
no single user responsible for REQM existed in the Ljubljana area.
Rather, there were many potential users of the information to be generated
in the study, so that a concerted effort was made to cultivate and involve potential users in the actual analysis. The lack of a single agency
as a focal point necessitated that the study goals and the study itself
be formulated in terms of demonstrating the utility of REQM analysis and
the various questions which REQM analysis can answer.

Given the above, the following were the goals of the study. One
goal was to attempt to apply the REQM approach to, and to test its utility in, a specific "real-world" context in which it was likely that many
data for analysis of REQM strategies did not exist. A second goal was
to assess the utility of applying, and the differences in so doing, a
methodological approach developed in the U.S. to another socio-economic-
political system. The hypothesis was that the methodology itself is
apolitical, the institutional context and constraints are not. A third
goal was to generate useful and meaningful information on the implications
of alternative REQM strategies for the specific area of Ljubljana, useful
in terms of actual decisions with respect to allocation of resources to
REQM in the Yugoslav context, meaningful in relation to Yugoslav organizations and decision makers. A fourth and final goal was to stimulate
Yugoslav agencies and organizations to take an active role in the study,
so that they could: (a) insure that the analysis was as Yugoslav-oriented

as possible; and (b) eventually expand upon and carry out the work initiated by the project.

Organization of the Report

Following this introductory chapter, chapter II describes the conceptual framework of REQM and defines terms used in subsequent chapters. Chapter III is a description of some of the physical, economic, political, and social characteristics of the study area. Chapter IV describes the actual planning of the analysis in terms of: (a) defining the conditions for the analysis; (b) selecting an analytical methodology; (c) defining study outputs; and (d) allocating available resources for the study. It is important to identify explicitly all the assumptions, constraints, and limitations relating to the analysis. Only by so doing is it possible to establish some bounds on the nature and validity of the decision-making information generated.

Chapters V and VI describe the analysis of activities which generate and discharge residuals to the environment in the area, and the development of cost estimates for achieving various levels of reduction in residuals discharges from those activities. Chapter VII explains the natural systems model used for estimating the impacts of residuals discharges on air quality in the area and what natural systems models need to be developed to make analogous estimates of impacts on water quality.

Chapter VIII describes how the information developed in the previous three chapters was used in developing, and analyzing the impacts of, alternative combinations of physical measures to achieve various environmental quality targets for the study area. The final chapter, IX: (a) describes how the physical measures analyzed for reducing discharges of residuals from the activities in the Ljubljana area were selected; (b) discusses some possible implementation incentives and institutional arrangements relating to the physical measures which might be applied in the Ljubljana area; (c) summarizes the major specific results of the study; and (d) discusses necessary next steps for REQM analysis in the Ljubljana area.

Chapter II

RESIDUALS, REQM, AND ANALYSIS FOR REQM

Key Definitions and Concepts

Before discussing the application of the REQM framework to the
Ljubljana area, some definitions and explanations of concepts are necessary. This chapter includes: the definition of a residual and a description of the interrelationships among types and forms of residuals;
the concept of an REQM system; definitions of REQM and of an REQM strategy; a discussion of analyzing and evaluating REQM strategies; and some
observations on the basic nature of REQM problems in all societies.

Residuals

All human activities--households, farming, manufacturing, mining,
logging, transportation--result in the generation of residuals. This is
because no production or use activity transforms all of the inputs to the
activity into desired products or services. The remaining flows of
materials and/or energy from the activity are termed nonproduct outputs.
If a nonproduct output has no value in existing markets or a value less
than the costs of collecting, processing, and transporting it for input
into the same or another activity, the nonproduct output is termed a
residual. Thus, residual is defined in an economic sense. Hence,
whether or not a nonproduct output is a residual depends on the relative
costs of alternative materials and/or energy which can be used instead
of the nonproduct output. These costs in turn depend on the level of

technology in the society at the point in time and on various governmental policies, both of which can change over time.

Figure 1 illustrates the definitions of residual, materials/energy recovery, and byproduct production. Some of the nonproduct outputs generated in an activity, particularly manufacturing activities, are recovered and are directly reused in the same production process (materials/energy recovery).

Depending on the relative costs of the alternative factor inputs, more or less recovery and reuse of the nonproduct outputs will be undertaken in the absence of constraints of one type or another on discharges of residuals to the environment. That is, materials and energy recovery will, in principle, take place up to the level where the marginal cost of an additional unit of recovered material or energy just equals the value of the recovered material or energy, as determined by the pricing system in the society. Because both the cost of recovery and the value of the recovered material or energy often change over time, the extent of recovery, and hence residuals generation, are likely to vary over time. The same principle holds for byproduct production, that is, where nonproduct outputs are used as inputs into a different production process, either at the same location as where the nonproduct outputs were generated or at another location.

The costs of recovery and of byproduct production and the values of recovered materials and/or energy and of byproducts often change over time. Therefore, the extent of recovery, the extent of byproduct pro-

Figure 1. Definition of Residuals

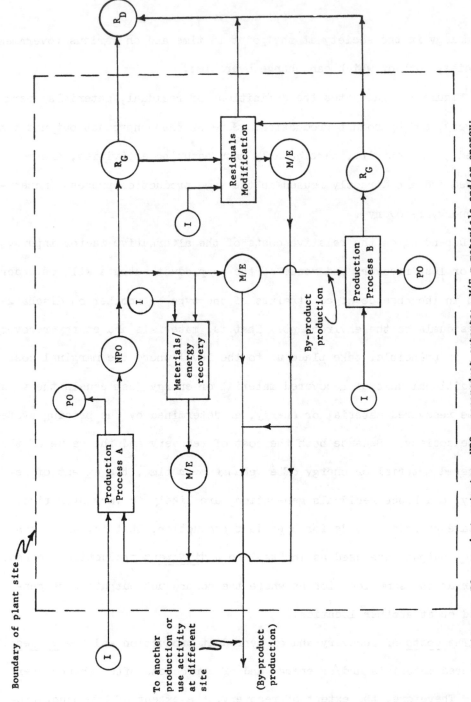

I = inputs; PO = product outputs; NPO = nonproduct outputs; M/E = materials and/or energy; R_G = residuals generated; R_D = residuals discharged

duction, and residuals generation are likely to vary over time. The remaining nonproduct outputs--the quantities remaining after <u>economical</u> recovery and byproduct production have been undertaken--are residuals.

Material and energy are the two basic classes of nonproduct outputs and residuals. The former occurs in the three states of matter: liquid, gaseous, and solid. The major energy residuals are heat and noise. Radioactive residuals have characteristics of both material and energy residuals.

<u>Interrelationships Among Residuals</u>. One form of material residual can be transformed into one or more other forms by the addition of materials and/or energy. For example, modification of sewage in a sewage treatment plant results in the generation of a semi-solid residual, sludge, plus various types of liquid residuals. If the sludge is incinerated, a gaseous residual, particulates, is generated. The material and energy inputs required for the modification themselves become residuals. Modification is undertaken under the assumption that the discharge of the modified residual and the discharge of residuals generated in its modification, will have fewer adverse impacts than the discharge of the original residual.

<u>Factors Affecting Residuals Generation</u>

Having defined residuals, the factors affecting residuals generation can be delineated. That is, what factors affect the types and quantities of residuals generated in any given activity--manufacturing, mining,

agriculture, transportation, households--per unit of activity, such as
per ton of product, per barrel of crude petroleum processed, per bushel
of wheat, per vehicle-kilometer, per capita, per household. In the
absence of controls on discharges of residuals, the residuals generated
in manufacturing, for example, are a function of: the characteristics of
the raw materials used; the technology of the production process, including age and physical arrangement of plant; the product mix; the specifications of each of the desired products; the operating rate, units of
raw material processed or units of output per unit of time; and prices
of factor inputs. Referring to figure 2, the residuals generated per ton
of paper towels produced are a function of: the species of wood used;
the method of wood preperation; the type of pulping process; the characteristics desired in the final product, such as wet strength, softness,
absorbency, and color; the type of paper machine; the number of tons of
paper produced per hour; and the prices of various factor inputs, such
as fuel, chemicals, wood, water, electric energy. The degree of whiteness
desired is of particular importance, because this determines the amount
of bleaching required and bleaching is a major source of residuals generation in the production of paper products. Shifting from white to unbleached paper products wherever possible, i.e., brown instead of white
towels, while holding all other product specifications constant, would
substantially reduce residuals generation.

Similarly, the residuals generated in producing wheat are a function
of such factors as: the type of soil; topography; type(s) of cultivation

19

Figure 2. Simplified Process Flow Chart for Production of Paper Towels

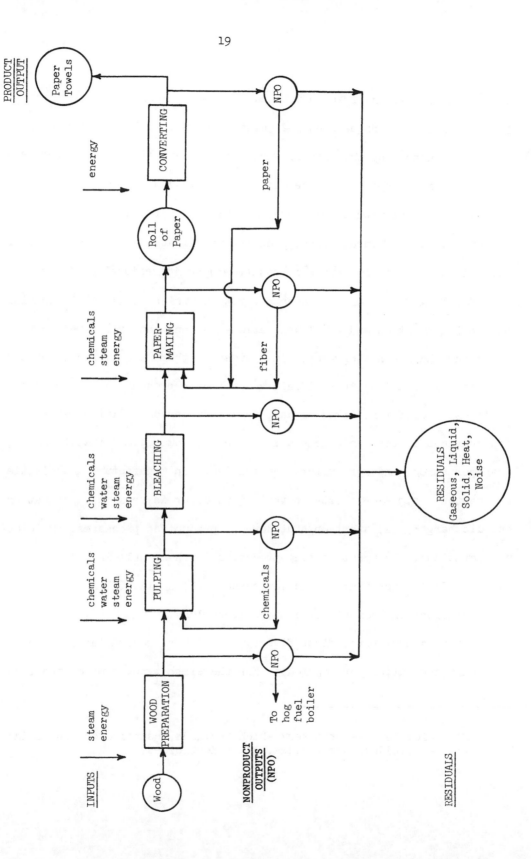

equipment; types, and frequencies and methods of application, of fertilizer; types, and frequencies and methods of application, of pesticides;[7]
prices of factor inputs, such as fuel, seeds, fertilizer, and water; and
climate. With respect to households, residuals generation is a function
of: design of structure; exposure; whether a single unit or multi-unit
structure; size of yard; income; education; climate; and prices of water,
electric energy, fuel, sewage disposal, and solid residuals disposal.

Rarely, if ever, is there an activity in which residuals generation--
per unit generation and total generation--is constant over time, in the
short-run and in the long-run. In residences there are diurnal, weekend,
seasonal fluctuations in residuals generation representing different activities at different times of day, week, and season. Similar short-run
fluctuations in residuals generation occur in commercial, institutional,
transportation, and agricultural activities. In manufacturing activities
at least six types of short-run variations occur in residuals generation
and discharge: (1) start-up/shut-down of production processes; (2) clean-
up operations; (3) upsets during production with no halt in production;
(4) breakdowns such that production ceases; (5) accidental spills; and
(6) variations in "normal" production operations.

The reasons for the first five types of short-run variations in
residuals generation are obvious. For the sixth type, some of the fac-

---

[7]Pesticide is a generic term which includes insecticides, fungicides,
herbicides, miticides, nematocides, and rodenticides.

tors which result in less than daily and daily variations in residuals generation and discharge are: (a) variations in quantity and quality of raw material inputs; (b) variations in operating conditions, e.g., operating a production unit beyond design capacity; (c) variations in the condition of operating equipment; and (d) daily and weekly variations in product mix.

Definite seasonal patterns exist for some production operations, e.g., in a large cannery packing various fruits, vegetables, soups, fish, where residuals generation depends on which raw product is in season; a petroleum refinery producing both gasoline and fuel oil, where the proportion of the two varies with the season (winter versus summer); the seasonal production cycle of automobile assembly plants. Many activities other than manufacturing exhibit major variations by season, such as agricultural operations, logging operations, resort operations. Household activities also exhibit substantial seasonal variation.

Over the longer run, several years or more, generation per unit is likely to change even if the level of output of the activity remains the same. Changing unit generation results from both endogenous and exogenous factors. Typically, for a given type of activity in a given facility--production of paper or steel or wheat--residuals generation per unit changes over time as a result of changes in endogenous factors such as production technology, aging of equipment, types of raw materials used, product mix, and product specifications. Exogenous factors affect residuals generation in an activity both directly and indirectly. Changes in prices of such factor inputs as fuel, water, electrical energy, chemicals, labor will usually result directly in changes in residuals generation by a given activity. Indirect effects stem from the fact that the value of a given material or energy nonproduct output generated by an activity is very likely to change over time in relation to the value of competing materials and energy for a particular use. An example is copper wire as a secondary (nonvirgin) source of metallic copper in relation to copper ore as a source of copper. As high quality copper ore deposits have been exhausted and the technology of processing copper wire to produce copper has become more efficient, the value of copper wire relative to copper ore has increased.

REQM System

The concept of an REQM system is illustrated in figure 3. Within any given region (however defined) at a given point in time, there is a spatial distribution of activities: industrial, mining, residential, agricultural, commercial, transportation. This spatial distribution of activities reflects the final demand for goods and services within the region and from outside the region. For each activity there are: (1) alternative combinations of factor inputs and related production technologies to produce the required goods and services, with a set of types and quantities of residuals generated associated with each combination; and (2) alternative ways of handling the residuals after generation. Activities as sources of residuals can be classified as: point sources, such as a power plant, a manufacturing plant, a residence; line sources, such as traffic flow on a major street; and dispersed or area sources, such as a logging area, an agricultural operation. Complex point sources, such as

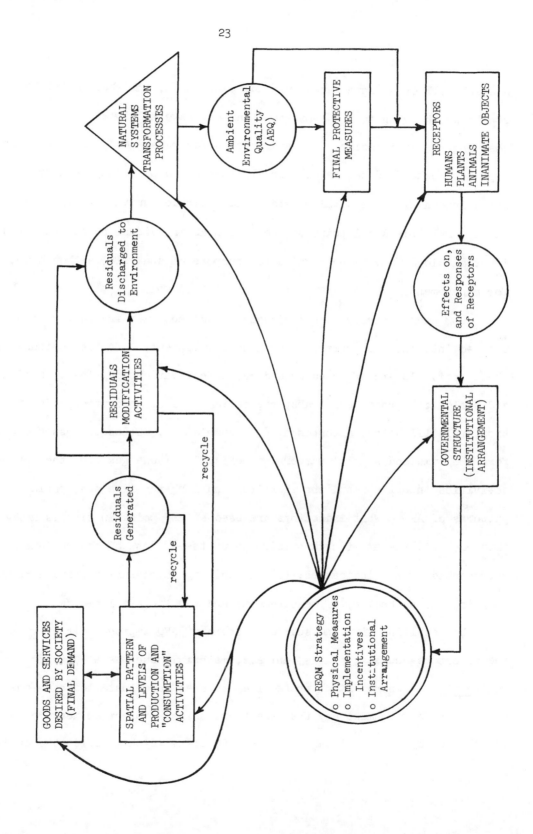

Figure 3. Concept of a Residuals-Environmental Quality Management (REQM) System

a steel mill or an integrated pulp and paper mill, will have multiple
stacks discharging to the atmosphere and may have multiple pipes discharging to water bodies. For analysis and/or management, each stack or
pipe may be handled separately, or all stacks and pipes may be aggregated into a single equivalent stack and a single equivalent pipe. Similarly, multiple small point sources, such as individual houses in a residential block, may be aggregated for purpose of analysis, although not
for management.

From each activity some residuals are directly or indirectly discharged into the air, water and/or land environments. In the environment
these residuals are affected by and may affect various physical, chemical,
and biological processes: transport, sedimentation, absorption, adsorp-
tion, volatilization, decomposition, accumulation. These processes transform the time and spatial pattern of residuals discharges from the various
activities into the resulting short-run and long-run time and spatial
patterns of AEQ. What indicators are used at any point in time is a function of: (1) the existing knowledge about the effects of the residuals
represented by the indicators; (2) the ability to measure the indicators,
that is, the technology of measurement; and (3) the available data.

The resulting time and spatial pattern of AEQ impinges _directly_ on
the receptors--humans, plants, animals, materials such as structures--or
indirectly, as where "final protective measures" are installed between
the ambient environment and the receptors. Final protective measures are
exemplified by a water intake treatment facility which modifies the qual-

ity of water withdrawn from a water body before the water is distributed for use and a filter on an air intake. The impacts on the receptors, as perceived by human beings, and the responses of individuals and groups to the perceived damages, provide the stimulus for action. The extent and form of action, as expressed in a selected REQM strategy, depends on the institutional structure, culture, and value system, and competing demands for scarce resources for other desired goods and services.

REQM

REQM consists of the following functions: analysis to develop REQM strategies; planning; legislation; translation of legislation into guidelines and procedures; implementation of guidelines and procedures via incentives imposed on residuals generating activities to induce those activities to install and operate physical measures for reducing the discharge of residuals into the environment and/or for modifying or making better use of the available assimilative capacity; design/construction/operation of facilities; monitoring and enforcement of performances by activities; monitoring of AEQ; and feedback of information from monitoring into the continuous planning and decision making functions. It is this total set of functions which yields the desired product of improved AEQ.

REQM on a day-to-day basis takes place primarily at the regional level. The region may be a metropolitan area, a river basin, an airshed, a soil conservation district, an economic region, or some combination of local jurisdictions. Experience indicates that it is less important that the boundaries of the region include all of the residuals dischargers

and all those affected by changes in AEQ, than it is that the boundaries
represent some region or area for which there is an institutional arrangement which can be made responsible for REQM. However, where cross-boundary
transfers of residuals and their effects are significant, they must be
explicitly considered. It is also extremely important to recognize that
factors such as national policies with respect to taxes, tariffs, imports,
and the prices of factor inputs established in national markets, affect
REQM in a region. Such factors are determined exogenously to the region,
and comprise part of the context in which REQM takes place and are major
factors which should be--but often have not been--explicitly considered.

REQM Strategy

An REQM strategy is comprised of: (1) the physical measures for improving AEQ; (2) the implementation incentives to induce the residuals
generators to apply the measures; and (3) the institutional arrangements
through which the implementation incentives are applied and the other related activities of REQM are carried out. These three interrelated components are depicted in figure 4.

Figure 4. Components of an REQM Strategy

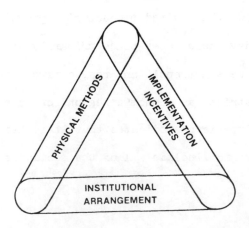

Physical Measures for Improving AEQ. The two basic classes of
physical measures for improving AEQ are shown in Table 2: (1) reducing
discharges of residuals to the environment; and (2) making better use of,
or increasing the assimilative capacity of, the environment. The first
class has two subclasses: (a) reducing generation of residuals; and (b)
modifying residuals after generation. Reduction in generation can be accomplished by changing the characteristics of goods and services desired,
changing production processes, and changing raw materials. Often these
three are interrelated, that is, a change in product specifications will
enable or require changes in both production processes and raw materials.
For a residence, changing plumbing fixtures and appliances represents
analogous changes to changes in production technology in a manufacturing
plant.

It is important to recognize that physical measures can be adopted
by a given activity on-site which will have no effect on residuals generation by the activity itself, but will reduce residuals generation elsewhere, in the same region or in other regions. For example, physical measures which result in reduced electrical energy use by an activity will
reduce the quantity of kilowat hours which have to be generated and hence
the associated residuals generated in producing the electrical energy.
Similarly, if the management of a center-city office building shifts from
bleached to unbleached paper towels--all other specifications of the
towels remaining the same--residuals generation in the office building
will not be changed but residuals generation at the loci of production of

Table 2. Classification of Physical Measures For Improving Ambient Environmental Quality

MEASURES FOR REDUCING THE DISCHARGE OF RESIDUALS

Category/Subcategory	Examples
Measures for reducing residuals generation	
Change raw material inputs	High to low sulfur oil in power plants; chips for round wood in pulp and paper production; pellets instead of run-of-mine ore in steel production; high to low or non-phosphate detergents; controlled release pesticides (in solvent and/or emulsified pesticide formulations) in agriculture
Change production process, including mode of transport	Individual vehicles to mass transit; conventional tillage to minimum or no tillage in agriculture; batch cooking to continuous cooking in food processing; conventional bleaching to oxygen bleaching in pulp production; open to covered feed lots in animal raising operations; solar heating for some portion of gas/oil heating in residential/commercial structures
Change mix of product outputs	Reduce number of products--types, grades, styles--produced in a given plant
Change product specifications	Reduce brightness of paper products, such as consumer products, writing paper, publication grade paper; reduce stringency of blemish standards for "grade A" fresh fruits and vegetables; decrease weight and motive power of automobiles; high-octane to low-octane gasoline; non-reusable to reusable containers
Measures for modifying residuals after generation	
Materials and/or energy recovery (direct recycle into same production process)	Reuse of mill scale in steel production; chemical and fiber recovery in pulp and paper production; reuse of cutting oils in machining operations
Byproduct production (indirect recycle, that is, to a _different_ production process)	
To intermediate products	Sulphite waste liquor to industrial yeast; obsolete vehicles into steel scrap; combustion of solid residuals to generate electrical energy; recirculation of household water from bathing and/or washing to toilets and/or lawn watering
To final products	Tomato pulp to pet food; citrus peels into candy; peach pits into charcoal briquettes; wood products residues into pressed logs; used rubber tires into artificial reefs and breakwaters
Modification without recovery or byproduct production	Neutralization, sedimentation, biological oxidation of liquid residuals streams; composting, incineration, land filling of solid residuals; electrostatic precipitation, wet scrubbing of gaseous residuals streams

Table 2. (Continued)

II. MEASURES DIRECTLY INVOLVING ENVIRONMENTAL ASSIMILATIVE CAPACITY

Category/Subcategory	Examples
Measures for making better use of assimilative capacity	
Change the temporal distribution of discharges from an activity	Withhold liquid residuals during periods of low assimilative capacity, e.g., impound liquid residuals from autumn canning season for release during winter and spring; store liquid residuals generated in precipitation events temporarily, for later gradual discharge, e.g., urban storm water runoff, feedlot runoff
Change the spatial location of residuals discharges from an activity	Pipe residuals to discharge locations where assimilative capacity is greater, e.g., gaseous residuals to stack on top of hill, liquid residuals from up-stream to downstream portion of an estuary
Change the time distribution of activities	Stagger office hours; restrict vehicle movements during certain times of day; use household appliances in off-peak periods; phase the initiation of new activities to coincide with the availability of residuals handling services
Change the spatial distribution of activities	Reroute vehicle movements; restrict density of residential housing to capacity of soil for on-site residuals disposal systems
Measures for increasing assimilative capacity	
Dilution	Add water from surface or ground water reservoirs to water bodies; reduce land elevation(s) around traffic junctions
Artificial aeration	Surface or subsurface diffusers to add oxygen to streams, lakes, estuaries
Artificial mixing	Mechanical mixing (destratification) of lakes and reservoirs; use multiple outlets at different elevations from reservoirs

the towels will be. Of course, the reverse is also possible. Physical measures adopted by an activity on-site may not increase on-site residuals generation but may increase residuals generation elsewhere. The greater the linkages among activities in a given region--in terms of material and energy flows among those activities--the more important this fact becomes for REQM in the region.

Modification of residuals after generation can be accomplished by: materials recovery (direct reuse in the same production process); by-product production or indirect reuse (use of a residual as an input to a different production process, at the same or a different location); and modification without recovery or byproduct production, traditionally termed "waste treatment". Byproduct production and modification without recovery can be carried out either on-site or in a "collective" facility, one which handles residuals generated at multiple locations. Examples of joint facilities are: a waste oil reclamation plant which processes used oil from many garages and gas stations; a plant for recovery of acid from used steel mill pickling liquor from several mills; a municipal incinerator, landfill, sewage treatment plant. Joint facilities usually permit achieving economies of scale, thus making economically feasible an activity which would not be so if it were undertaken at each individual operation. As noted previously, residuals modification without recovery or byproduct production does not reduce the total quantity of residuals discharged into the environment, but rather merely transforms one type of residual into other forms of the same type and/or one or more other types.

In order to make the transformation, additional inputs are required, and hence additional residuals are generated.

The second basic class of physical measures for improving AEQ also has two subclasses: (a) making better use of existing assimilative capacity; and (b) increasing assimilative capacity. The former reflects the fact that assimilative capacity varies both over space and over time, that is, diurnally, seasonally, year-to-year. Examples of the former are: (a) building stacks for discharging gaseous residuals at levels such that greater dispersion can be achieved, to reduce ground level concentrations in the region;[8] (b) relocating outfalls to where there is greater assimilative capacity in streams and estuaries; (c) temporary withholding of liquid residuals during periods of low streamflow for subsequent discharge when streamflow, and hence assimilative capacity, is higher; (d) rescheduling of activities in time, such as changing the time pattern of traffic flow; and (e) relocating activities in space to be more in accord with the natural assimilative capacity in an area, such as locating industrial activities downwind from residential areas. (a), (b), and (c) involve no change in residuals generation per unit; (d) and (e) may change residuals generation, as when rerouting of traffic results in increasing average velocity, which in turn means fewer residuals generated per vehicle kilometer traveled. Increasing the assimilative capacity can be

---

[8]The result may be to increase ground level concentrations <u>outside</u> the region.

accomplished by: the addition of water to streams during low flow periods; artifically adding oxygen to streams, lakes, and estuaries; and building topography with solid residuals, e.g., making artificial hills.

Implementation Incentives. Implementation incentives are the inducements which stimulate the residuals generators and the REQM agencies to install, operate, and maintain the physical measures for improving AEQ. An implementation incentive can be positive or negative (reward or punishment), direct or indirect, prescriptive or proscriptive. Some implementation incentives can be applied at all levels of government--communal, republic, national--separately or simultaneously; others are more appropriately applied at one of the levels. Table 3 provides a classification of implementation incentives.

Some amplification with respect to implementation incentives is necessary. First, the separation of regulatory and administration implementation incentives is operational but arbitrary. Regulatory refers to imposition by one governmental agency on another governmental agency or on an enterprise; administrative means within a governmental agency or an enterprise. Some of the economic incentives, such as effluent charges and surcharges on energy use, can be, and have been, applied within single industrial plants.

Second, implementation incentives imposed on one residual can have positive and/or negative impacts on other residuals and/or environmental media. An ordinance specifying limits on concentration of particulate discharges from building incinerators may result in closing down some

Table 3. Classification of Implementation Incentives

REGULATORY--by law, ordinance, permit

Specification of a physical measure

    Specify characteristic(s) of raw material input, e.g., no more than 1% sulfur fuel, target-specific degradable pesticides

    Specify "production process", e.g., amount of thermal insulation in buildings, contour plowing in agricultural crop production, dry peeling in fruit and vegetable canning

    Specify residuals modification and/or handling process, e.g., activated sludge sewage treatment plant, debris basins on construction sites, require householders to separate used newspapers and paper packaging from other solid residuals

    Specify product output characteristics, e.g., amount of lead in gasoline, amount of phosphate in detergents, standard number/sizes/designs of glass containers

Specification of a result or performance

    Specify total quantity of residual per unit $\leq$ a specified amount, e.g., $\leq$ X kilograms of suspended solids per ton of steel, $\leq$ y grams of hydrocarbons per vehicle-kilometer traveled

    Specify total quantity of residual discharged per unit of time $\leq$ a specified amount, e.g., z kilograms of suspended solids per day, T kilograms of particulates per day

    Specify concentration of residuals in discharge $\leq$ a specified amount, e.g., 30 mg/l of suspended solids in waste water

    Specify AEQ level to be achieved, e.g., $>$ 6 mg/l dissolved oxygen in spring and fall, 5 mg/l remainder of year; mean annual concentration of sulfur dioxide $\leq$ 150 micrograms/m^3

    Specify performance, e.g., 85% BOD_5 removal (from specified base) in sewage treatment plant, 98% particulates removal (from specified base) in power plant stack, efficiency of energy use by appliance must be $\leq$ specified level, automobile must achieve at least 40 kilometers per liter of fuel in city driving

Specification of locations or limitations on locations of activities ,
e.g., designate industrial districts, limit density of construction where infiltration capacity for septic tanks is less than a specified rate, prohibit surface mining on slopes $>$ 25%

Specifications of extent, timing, type of activity, e.g., prohibit trucks on
particular routes during particular time periods, prohibit all-terrain vehicles in environmentally fragile areas, prohibit aerial spraying of pesticides when wind $>$ 8 kilometers per hour, stagger work hours

Specification of procedure, e.g., require environmental impact statement be pre-
pared on each project according to specified guidelines, semiannual testing and inspection by communal (republic) agency of discharge controls on motor vehicles, require public hearings before REQM strategy is adopted

Table 3. (Continued)

ECONOMIC

> Applied directly to residuals, e.g., charge on each unit of residual discharged
> e.g., X New Dinars (N.D.) per kg. of BOD_5, Y N.D. per 10^6 kilocalories, Z
> N.D. per kg. of SO_2

> Applied to inputs, e.g., surcharge on each unit of energy intake, charge on
> each kilogram of DDT applied

> Applied to product outputs, e.g., charge on each kilogram of packaging, annual
> surcharge on horsepower of automobile, severance taxes on virgin materials
> extracted (M N.D. per ton of coal mined)

> Applied to activities, e.g., reduced parking fees for car pools, parking sur-
> charges

> Applied to residuals modification, e.g., sewer and landfill user charges,
> reduced taxes for installation of residuals discharge reduction measures,
> republic grants to municipalities to finance installation and operation of
> residuals handling and modification facilities

ADMINISTRATIVE--by order within a governmental agency or an enterprise

> Applied directly to residuals, e.g., require separation of various types of
> paper residuals in offices, collect all used lubricating oil and used tires
> for reprocessing

> Applied to products used, e.g., specify that only unbleached consumer paper
> products can be used, specify that only raw materials in reusable containers
> can be purchased

> Applied to activities, e.g., specify limits on thermostat settings for heating
> and air conditioning

JUDICIAL

> Court or administrative law review and action, or threat thereof, to compel
> compliance, civil and/or criminal suits

EDUCATIONAL/INFORMATIONAL

> Educational programs to acquaint individuals/groups of the implications of their
> activities with respect to residuals generation/discharge and adverse impacts
> on AEQ and to acquaint them with alternative behavior patterns which would
> reduce such impacts

> Informational programs: appliance labeling with respect to energy efficiency,
> fuel economy rating, pesticide labeling

Source: Modified from B.T. Bower, C.N. Ehler, and A.V. Kneese, "Incentives for
Managing the Environment," Environmental Science and Technology vol. 11,
no. 3, 1977, pp. 250-254.

or many of these incinerators with a consequent increase in solid residuals
for disposal. The response to an effluent charge on BOD_5 discharge may
often result in simultaneous reductions in discharges of TSS and/or phenols.

Third, mixes of implementation incentives can be applied to a given
residual generated by an activity. For example, an effluent standard can
be coupled with an effluent charge, e.g., an upper limit of 400 kilograms
of BOD_5 discharge per day with 5 New Dinars (N.D.) per kilogram for each
kilogram in excess of 400 per day.[9]

Fourth, many implementation incentives for improving AEQ are not under
the jurisdiction of agencies responsible for REQM. Examples include: tax
policies such as depletion allowances, severance taxes, capital gains provisions, accelerated depreciation; import restrictions; tariffs; and land
use restrictions.

Fifth, some materials are so toxic that their discharge to the environment should simply be prohibited. However, it is no easy task to determine just what materials in what quantities under what conditions are
toxic to what species, homo sapiens included. A particular concentration
of a material in a body of water may have no adverse effect on some fish
species under low or normal temperature conditions, but very adverse effects under high temperatures, or in the presence of some other material,
or when dissolved oxygen is low. In some cases, other contributing or in-

---

[9]Many variations are of course possible, e.g., upper limit of 400 kilograms of BOD_5 discharge permitted per day with 1 N.D. per kilogram up to
400 kilograms, 10 N.D. per kilogram for each kilogram in excess of 400.
A major objective in selecting implementation incentives is to reduce the
upper end of the distribution curve of residuals discharges.

tervening variables may affect the response, such as the state of health
of the target species. Finally, toxicity is both a short-run and a long-
run problem. Concentrations that result in immediate fish kills are obvious; those that result in a build up in tissues and/or organs over time,
with long-term effects on reproduction and viability, are not so obvious.

A related problem is the lack of, or incomplete, knowledge of both the effects of the discharge of a residual on AEQ and the effects on various
species from the changes in AEQ. Again, both short-run and long-run time
frames are involved. The impact of discharges of fluorochlorohydrocarbons
on ozone content of the stratosphere is an example. The long-run impact
of particulate matter and CO_2 discharges on the atmosphere is another.

Finally, the impact of the implementation incentive on technological
change is important. The relevant question is, "Does the implementation
incentive bias the selection toward a particular technology for improving
AEQ, or does it stimulate the investigation of the entire range of alternative physical measures in relation to residuals handling technology, al-
tering factor inputs, changing product output specifications and 'life
styles', and changing production technology?"

Each implementation incentive has its strengths and weaknesses, so
that the implementation incentive(s) chosen in a particular REQM context
must be matched to that situation--specific residuals, specific activities,
specific physical measures. No one implementation incentive is likely to
provide optimal REQM at the communal, regional, republic, and national
scales.

Institutional Arrangement. The institutional arrangement for REQM is defined as the set of one or more institutions which has or can obtain the legal authority to impose implementation incentives on residuals generators and to carry out the collective tasks of REQM. Factors affecting the choice of institutional arrangement include the spatial dimensions of the REQM problem, available implementation incentives, legal considerations, political considerations, and the cultural values of the society. Consideration of the institutional arrangement is often the weakest link in the development of REQM strategies.

Analysis for REQM

Whatever the societal and institutional context, making decisions on how to improve AEQ requires information on the costs and consequences of alternative strategies. The rigorous development of such information is the analysis function of REQM. If planning is defined as the process of choosing an REQM strategy--the decision concerning what resources to allocate where, how, and when to produce what levels of AEQ, where AEQ is defined by a set of AEQ indicators such as fish biomass, dissolved oxygen concentration, SO_2 concentration, and number of acres devoted to landfills--then analysis is that activity which generates the input to this decision. The output of the analysis consists of the costs and consequences of alternative REQM strategies, criteria for evaluating those strategies, and the results of applying the criteria to the strategies.

Because the nature of the REQM problem(s) and the resources available for analysis--time, personnel, computational facilities, data--vary from region to region, the degree of sophistication used in the analysis should be "tailored" to the situation. Developing an REQM strategy is not a process which occurs in a vacuum. Rather, the procedure usually involves some region for which one or more REQM problems have been perceived or identified or within which questions have been raised about REQM. Given such a region the analysis requires: (1) the development of models of residuals generation and modification activities, termed "activity models"; (2) the development of models of the processes which affect, and are affected by, the residuals after their discharge into the environmental media, termed "natural systems models"; (3) the specification of an explicit objective function, which includes AEQ indicators either in the function itself or as constraints; (4) the selection of a method of analysis; and (5) the development and application of criteria for evaluating strategies.

Activity Models

For each major residuals-generating activity or group of activities-- manufacturing plant, household, municipal incinerator, farm, transport system--an activity model of greater or lesser sophistication is necessary. Such a model: (1) indicates alternative combinations of factor inputs to produce the given outputs of products and/or services; (2) delineates, for different sets of prices of those factor inputs, the types and quantities of residuals generated per unit of activity--ton of steel produced, kilowatt hour of electrical energy generated, bushel of corn produced, barrel of crude petroleum processed, inhabitant in household; and (3) identifies the various physical measures available for reducing the discharge of residuals into the environment and the costs of various degrees of discharge reduction. Depending on the time and resources available and on the relative importance of the different generators for the given REQM context, activity models may be highly aggregated and simple, or highly disaggregated and complex, and may take the form of mathematical models.

Natural Systems Models

The outputs of the activity models--types and quantities of residuals discharged at specific locations and times--comprise the inputs into the natural systems models, along with the relevant hydrologic, geomorphologic, meteorologic, and pedologic variables, such as temperatures, wind velocity, precipitation, soil characteristics, topographic slope, steam channel characteristics, and sunlight. Major types of natural systems models are: (1) physical dispersion models such as for suspended particulates, SO_2, and total dissolved solids (salinity); (2) chemico-physical dispersion models such as for photochemical smog, and pesticide movement and modification in ground water aquifers; and (3) biological systems models such as terrestrial and aquatic ecosystem models. Just as for activity models, the time and resources available and the relative importance of the environmental media--in terms of the relevant AEQ problems--determine the degree of complexity necessary for the natural systems models. For example, a water quality model may consist of a set of simple linear transfer coefficients or it may be a multi-compartment aquatic ecosystem model. Natural systems models transform the time and spatial pattern of residuals discharges into the environment into the resulting time and spatial pattern of AEQ, as measured by whatever indicators are of interest in the particular situation.

The linkage between activity models and natural systems models merits emphasis. The activity models must be formulated to provide the data inputs in the proper form for the natural systems models. Conversely, the latter must be formulated in relation to the outputs which the activity models can provide, the available data, and the desired outputs in terms of indicators of AEQ.

Objective Function and Method of Analysis

An objective function is a statement of the criterion or the criteria for which the best solution is desired. Constraints are additional factors which must be taken into account in the solution of the problem, factors which limit the range of permissible solutions. Criteria can be reflected in the specification of constraints as well as in the specification of the objective function. Constraints are also part of the decision making criteria. Specification of an explicit objective function and the accompanying constraints serve to translate the goals and/or specific objectives of a study into a mathematical or quantitative description of the REQM problem(s).

If damage functions are available which translate AEQ into monetary damages to receptors, such as costs associated with impacts on human

health, value of yield loss for agricultural crops, costs of cleaning
buildings, then the objective function can be expressed as maximizing the
present value of net benefits, that is, the present value of the time
stream of damages reduced minus the present value of the time stream of
REQM costs, over some specified time horizon with the relevant social rate
of discount.

Developing damage functions involves the determination of the relationship between short-run and/or long-run exposure to the concentration
of a residual and physical or physiological effects, and then the determination of the relationship between those effects and monetary costs.
Such determinations are difficult for at least three reasons. One, most
damage functions involve multiple variables in addition to the residuals
concentration-duration variable. For example, the effect on alfalfa of a
given concentration of any one of several gaseous residuals is higher at
high humidities than at low humidities. The impacts on human health of
SO_2, particulates, and probably other gaseous residuals depend on other
variables related to the receptor, such as general physical condition, age,
nutrition, and smoking habits. Two, often several residuals of the same
form or in a single environmental medium are simultaneously involved, such
as SO_2, particulates, and nitrogen oxides in the atmosphere, so that there
may be additive or mitigating effects. Three, the same residual, such as
lead, can affect the individual through ingestion into lungs in gaseous
form, through liquid intake, and in foods. Determining the relative impacts of each on the "total body burden" is difficult.

Generally, all the necessary damage functions are not available to enable utilization of a net benefit objective function in analysis for REQM. Recourse must then be made to some form of cost minimization objective function, such as minimize the present value of costs to meet a set of AEQ standards and/or limits on discharges, with or without a capital cost or capital plus operation and maintenance cost constraint, and/or with other constraints.

Once the objective function has been specified, two basic methods of analysis are available. One involves the use of mathematical programming, such as linear programming. Given: (1) a level of output for each activity; (2) the costs of different degrees of individual and collective residuals modification activities; and (3) the costs of different degrees of direct modification of AEQ, such programming specifies the minimum cost combination of physical measures to achieve the specified AEQ standards and/or to meet specified upper limits on residuals discharges.

The other method can be characterized as a manual "search" method. Given the specified level of output for each activity, one or more discrete options for reducing residuals discharges and the associated costs are delineated for each activity, including collective residuals handling and modification activities. As with the programming method, costs of different degrees of direct modification of AEQ are also delineated. Various combinations of these options are then specified, and the effects on residuals discharges, on AEQ, and on REQM costs are estimated and tabulated. The analysis of possible combinations is continued as long as computational resources are available. Rarely are sufficient resources available to make a search of all possible combinations. However, there are formal mathematical techniques which can be used to sample the response surface.[10]

Both methods of analysis enable investigation of the effects of the following variables on residuals generation and hence on AEQ in the region and on REQM costs: alternative spatial patterns of activities; alternative final demands; alternative transportation systems; alternative collective residuals handling and disposal facilities; and alternative AEQ standards and discharge constraints.

Criteria for Evaluating REQM Strategies

In REQM, as in all decision making contexts, criteria must be established by which to choose an REQM strategy. These criteria represent factors which decision makers consider relevant in evaluating strategies. Not only must the criteria be specified but relative weights must be attached to them.

Although real resource costs of an REQM strategy represent a major factor in choosing a strategy, these costs are not the only criterion. Decision makers in all societies use multiple criteria in making decisions, with the criteria and their relative weights (importance) being made more

---

[10]See Maynard M. Hufschmidt, "Analysis by Simulation: Examination of Response Surface," in A.A. Maass, et al, Design of Water-Resource Systems (Cambridge, Mass: Harvard University Press, 1962).

or less explicit. Table 4 lists a set of criteria for evaluating REQM strategies. These criteria may be applied to each residual/activity/ physical measure/implementation incentive combination, or to each REQM strategy as a whole. Although most of the criteria are self-explanatory, a few clarifying comments are warranted.

1. The changes in AEQ can result in physical effects on various receptors, such as decreased human morbidity and mortality, decreased deterioration of materials, increased fish populations.

2. Direct benefits represent the monetary value of the physical effects stemming from changes in AEQ, such as reduced medical costs, reduced costs of cleaning and maintaining structures, and values of increased harvests of agricultural crops and fish. Public and private administrative costs for accounting and reporting, monitoring, analysis of samples, supervision of operating personnel, inspection are components of direct costs, but they are separately identified because they are too often neglected. In many contexts another important consideration with respect to costs is the extent of foreign exchange required by the given strategy.

A very important consideration with respect to both physical and economic effects is their distribution. Who benefits from improved AEQ and who pays in what forms and over what periods of time for that improvement? Distributional effects should be determined in relation to: (a) political jurisdictions and socioeconomic groups of the population within the REQM region; and (b) the division between direct costs incurred within the region and those incurred external to the region.

Table 4. Criteria for Evaluating REQM Strategies

1. Physical Effects

    a. Reduction in discharge of residual from the activity
    b. Reduction in total discharge of residual in the region
    c. Changes in ambient environmental quality and their distributions
    d. Effects of changes in ambient environmental quality and their distributions

2. Economic Effects and Their Distributions

    a. Direct benefits
    b. Direct costs
    c. Administrative costs
    d. Indirect costs

3. Administrative Considerations

    a. Simplicity
    b. Flexibility

        (1) Retention of effectiveness under changing conditions
        (2) Continuous or noncontinuous application
        (3) Selective or uniform application

4. Timing Considerations

    a. Years before physical measure is in place and is operating
    b. Years before impact on ambient environmental quality is realized

5. Political Considerations

    a. Priority in relation to other environmental quality management problems
    b. Priority in relation to other societal problems
    c. Impact on intergovernmental relations, i.e., intercommunal, communal-republic, communal-federal, republic-federal
    d. Acceptability to public
    e. Legal difficulties

6. Intermedia and Resource Use Effects

    a. Intermedia effects: quantities of other residuals generated-- gaseous, liquid, solid, noise, thermal
    b. Resource use effects: net energy use, net land required, net consumptive use of water

7. Accuracy of Estimates

    a. Physical effects
    b. Costs

Source: Modified from B.T. Bower, C.N. Ehler, and A. V. Kneese, "Incentives for Managing the Environment," _Environmental Science and Technology_ vol. 11, no. 3, 1977, pp. 250-254.

3. Flexibility in administration refers to the degree to which a strategy remains effective under changing conditions and to the ease with which the strategy (or elements of it) may be applied or removed. Changing conditions include: (a) seasonal and short-term variations in both residuals generation and assimilative capacity; (b) changes over time in factor prices, such as fuel, energy, water, and changes in technology; (c) new information obtained over time with respect to the behavior of the natural systems involved and the behavior of residuals generators in responding to implementation incentives and to changing factor prices; and (d) new social goals and priorities, i.e., as society's tastes change. The continuous/noncontinuous attribute of flexibility refers to whether a physical measure/implementation incentive combination is applied continuously or can be applied intermittently as needed. The selective/uniform attribute refers to whether the physical measure/implementation incentive combination is applied to all activities or can be applied to selected activities. Combinations of the continuous/noncontinuous, selective/uniform attributes of REQM strategies are shown below. REQM costs are

Timing of Application

		Continuous	Noncontinuous (Intermittent)
	Uniform		
Coverage/ Selectivity/ Loci of Application			
	Selective		

almost always highest with continuous, uniform application.

4. Timing considerations relate to the fact that physical measure/
implementation incentive combinations vary with respect to both the time
required to install the physical measure and place it in operation and the
time required after it is in operation before the effect on AEQ occurs.
Timing may be a particularly important consideration where there are adverse AEQ conditions which need to be ameliorated as soon as possible.

5. The political considerations criterion has five components. The
first refers to the perceived urgency of the particular REQM problem in
relation to other REQM problems, for example, improved air quality compared
to improved water quality, or in relation to other locations of the same
REQM problem. The second refers to the perceived urgency of REQM problems
in relation to other societal problems in the region. The third is the
impact on intergovernmental relations, in terms of the strategy's effect
on the normal way of carrying out government business. The fourth component
is public acceptance. An REQM strategy which is new and/or unexplained to
the public may inhibit acceptance. Presumably public acceptance is most
easily gained through the involvement of the public from the initiation of,
and throughout, the planning process. The fifth component relates to the
degree of difficulty in obtaining legal authority for the institutional
arrangement to implement the strategy. Does adequate authority to implement the strategy exist; could existing legislation be changed to enable
implementation; would entirely new legislation have to be passed or exec-
utive decrees promulgated?

6. Intermedia and Resource Use Effects should be tabulated in physical terms, as well as being incorporated in economic costs. The three
resource use effects most often of interest are: net energy required, net
land required, and net consumptive use of water. An REQM strategy may be
energy intensive, or it may actually reduce total energy use in the region.
This is an important factor where a significant proportion of the basic
energy supply, or of a particular type of fuel--such as petroleum--must be
imported. The use of ponds, lagoons, and/or spray irrigation to reduce
the discharge of liquid residuals may increase the net consumptive use of
water in a region. The land required by an REQM strategy, for example for
disposal of mixed solid residuals and sludge, may be an important consideration in a densely urbanized area, and may be inadequately reflected
in the estimation of land costs.

7. Accuracy of the estimates of the costs of the REQM strategy and of
the impacts on AEQ which the strategy is predicted to have may affect the
choice of strategy. A strategy which has large estimated positive effects
on AEQ and/or low costs, but for which there is large uncertainty in the
estimates of costs or effects or both, may not be preferred to one which
has substantially less impact on AEQ and higher costs but for which the
probability of achieving those effects is high.

After evaluating each strategy according to each of the indicated criteria, the final step in evaluating strategies is to combine the ratings.
This process involves assigning relative weights to the criteria, an activity which is the responsibility of the decision makers, not of the an-

alysts, and selecting either an additive or a multiplicative method of combining the weighted criteria.

Concluding Comments: Some Basic Considerations in REQM and Analysis for REQM

The preceding discussion has described   the pervasiveness of and the factors determining residuals generation, the concept of an REQM system, the role of the three environmental media in providing residuals assimilation services, the three components of REQM strategies, and analysis for REQM. The following are basic considerations which all REQM managers and analysts should take into account explicitly.

First, the crux of the residuals problem stems from the magnitude and proliferation of goods and services "demanded" by society. This has three facets: longevity, style, variety. To produce a specified set of production and so-called consumption goods, the quantity of residuals generated decreases as the useful lifetime of the goods increases. The longer machinery, cars, buildings, and other durables last, the fewer new materials are required to compensate for depreciation and/or to sustain a given rate of capital accumulation, and fewer residuals are generated in total. The same is true for energy production, the more efficient the energy generation/conversion processes, the fewer residuals will be generated for a given level of energy output. The impact of style changes, as in annual automobile and clothing changes, is to encourage "artificial" obsolescence and reduce longevity. Increasing the multiplicity of models of container sizes, automobiles, cosmetics increases total materials and energy throughout for a given level of output, thereby increasing residuals generation.

Second, an obvious corollary of the first is that REQM costs depend to a very significant degree on consumer choices and business decisions. Thus, one of the most important questions is, how can a society stimulate the development of production processes, product outputs, and services which will reduce the generation of residuals in the first place, and induce consumers to choose (prefer) those products and services?

Third, strategies for REQM, as with decisions regarding programs in other sectors of an economy, are determined in the institutional milieu of society. Therefore, an important element affecting choice of strategy involves the perceptions of, and attitudes toward, various aspects of AEQ. Perceptions and attitudes of individuals and groups are particularly important because all damages from adverse AEQ and benefits from improving AEQ cannot as yet be measured in monetary terms. Thus, to many men and women "on the street," the visible plume from a stack connotes air pollution, even though the plume may be only water vapor. Just what is "clean air" and "clean water" in the minds of people, politicians, different groups in society? Who is, or which societal groups are, concerned about AEQ?

Fourth, AEQ varies both over time and over space. AEQ is stochastic because of the basically random variation of assimilative capacity over time and the similar random variation in generation and discharge of residuals by activities in the region. Assimilative capacity varies from location to location within a region and from region to region. Studies have shown that:

a. failure to consider the locations of liquid residuals discharges in a given region can increase costs by 50 percent or more to achieve a desired level of ambient water quality, that is, if all dischargers are required to reduce by the same amounts regardless of their impacts on water quality;

b. failure to consider the locations of gaseous residuals discharges and the affected receptors in an airshed--in relation to variation in as-similative capacity over the airshed--can at least double the costs to achieve a given level of ambient air quality;

c. failure to consider that "harmful" or undesired levels of air quality in many airsheds occur during less than 5 percent of the days, while at the same time requiring control measures on gaseous residuals 100 percent of the time, can result in quadrupling costs, to achieve a given level of ambient air quality; and

d. failure to consider variations in assimilative capacity of water bodies among regions, that is, requiring uniform discharge reduction, will about double national costs to achieve a given level of ambient water qual-ity.[11]

Therefore, the analysis of REQM must include explicit consideration of the stochastic worlds of the air, water, land environments, and residuals

---

[11]R.A. Luken, D.J. Basta, E.H. Pechan, The National Residuals Discharge Inventory (Washington, D.C.: National Research Council, January 1976). Also available through the National Technical Information Service, U.S. Department of Commerce as PB-252-288.

generators and explicit consideration of space.

Fifth, the economic feasibility of REQM strategies is heavily dependent on the spatial distribution of activities at any point in time and over time. This is particularly true in urban areas. For example, transportation predominantly by mass transit--electrically powered or otherwise--will rarely be feasible except where the urban pattern consists of a core and concentrated subcenters. Similarly, heating of dwelling units from communal rather than individual systems is not feasible with a dispersed settlement pattern. The costs of many of the collective residuals management measures, from sewage collection systems to residuals recovery-recycling systems, increase non-linearly and rapidly with urban sprawl.

Chapter III

THE LJUBLJANA AREA

This chapter describes the physical and economic characteristics of the five-commune area defined for the REQM analysis, the institutions involved in REQM in the area, and the factors considered in demarcating the area. Following this profile, Chapter IV describes the process of setting up the REQM analysis for the area.

Location and Characteristics of the Five-Commune Area

The five-commune area is located in Slovenia, the most northerly and most westerly republic of Yugoslavia, as shown in figure 5. The area is situated in a basin along the southern slopes of the Julian Alps. The five communes comprising the area are Centar, Siska, Vic, Moste, and Bezigrad, covering 903 square kilometers (350 square miles).[12] The five communes comprise about 90 percent of the basin's land area. In the center of the basin is the city of Ljubljana and its suburbs, designated as the urbanized area in figure 6. The urbanized area--comprised of all of Centar commune and parts of the other four communes--includes about 5 percent of the land area of the basin, and contained, in 1972, about 70 percent of the population of the five communes. To the south of the city there is a large area of poorly drained, high water table land, known as

---

[12]Zelena Kniga o Ogrozenosti Okolja v Slovenijil (The Green Book of Environmental Pollution in Slovenia) (Ljubljana, Yugoslavia: Zavod za Statistiko, SR Slovenije, 1971).

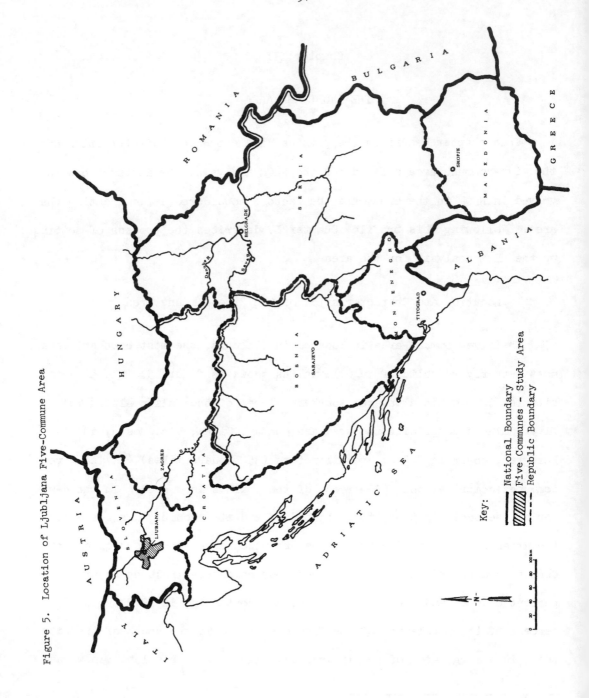

54

Figure 5. Location of Ljubljana Five-Commune Area

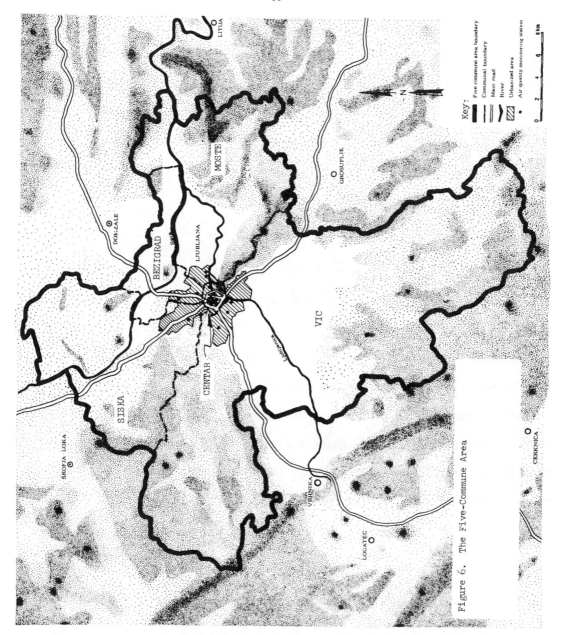

Figure 6. The Five-Commune Area

the moor area; to the north lies a dry alluvial plain in which the remaining 30 percent of the population of the five communes resides. The
mountains bordering the basin are a major center for alpine sports.
Much of the population engages in such activities as skiing and hiking
and consequently is very much aware of the value of the natural resources
making these activities possible.

Air Resources

The meteorological characteristics of the basin result in a nearly
closed system with respect to air quality. The general climate of the
area is classified as temperate. Periods of fog, temperature inversion,
and limited ventilation occur which are most severe in winter and often
last seven days, and occasionally longer. The ambient quality of the
relatively stationary air masses during such periods is often poor, because of the accumulation of gaseous residuals discharged from activities
within the basin. Air quality within the basin does not appear to be
affected significantly by imports of gaseous residuals into the basin.

Water Resources

Both ground and surface waters are used for water supply in the area.
The water for the Ljubljana city system is withdrawn from ground water
aquifers in the recharge area north of the city. Most of the water which
is used in industrial operations, other than for potable water, is withdrawn from local surface sources. Users outside the area supplied by the
city water system depend on individual wells tapping the ground water
aquifers.

Two major rivers flow through the basin. The Sava, the largest river in Yugoslavia, flows in a southeasterly direction in Slovenia through most of the country to its confluence with the Danube at Beograd. Just east of Ljubljana the Sava is joined by the Ljubljanica, which flows eastward through the center of the city. Most of the liquid residuals from the activities in the study area are discharged unmodified into these two rivers, or into one of the several smaller tributaries. Water quality of the Ljubljanica is very poor. However, because of the rapid rate of flow and substantial reaeration, water quality of the Sava in the five-commune area is generally acceptable.

Employment

In 1972, the population of the five-commune area was about 257,000,[13] an increase of 25 percent over the 1962 population. During this period a significant shift occurred in the mix of economic activities in the area. Major new production activities initiated in the 1962-1972 period included industrial--primarily metal processing, electrical products, and food processing--and commerce, including a significant amount of tourism in the summer. Agriculture declined in importance relative to other activities.[14] In addition, much of the agricultural activity is private and

---

[13]*Statistical Letopis SR Slovenije* (Statistical Yearbook for Slovenia) (Ljubljana, Yugoslavia: Zavod za Statistiko, SR Slovenije, 1974).

[14]In 1972 only about 1.9 percent of the gross domestic product of the five-commune area was attributable to agriculture; the percentage for the nation as a whole was 18.8 percent. See *Yugoslav Survey*, vol. XV, no. 1 (Beograd, Yugoslavia: Jugo Slovenski Pregled, Feb. 1974).

unmechanized, with little use of chemical fertilizers and pesticides.

The 1972 work force of about 105,000 individuals was distributed as shown in table 5. About 14,000 of these are commuters who live within the five-commune area outside of the urbanized area and commute to jobs within the urbanized area.

Institutional Arrangements

As stated in chapter I, the responsibility for environmental quality management in Yugoslavia is delegated to the communes. At that level of government no single agency exists with responsibility for environmental quality decision making or for REQM. However, there are four institutions which can influence REQM strategies in the five-commune area.

One is the Environmental Council of the city (urbanized area). This is a non-technical advisory group which provides advice on environmental issues to the city assembly. A second is the Republic Committee for Air Quality, which is an advisory group of professional and technical people. The function of the Committee is to formulate legislation at the republic level relating to air quality management. The other two, mentioned in chapter I, are Splosna Vodna Skupmost, the technical arm of the Slovenian Republic Water Quality Commission for the Upper Sava River Basin, and SEPO, a non-governmental agency affiliated with Ljubljanska Banka.[15]

---

[15]Ljubljanska Banka is the largest bank in Slovenia. At the time of the study it was financing about 70 percent of the investments in the Republic.

Table 5. 1972 Employment in Major Activities in the Five-Commune Area[a]

	INDUSTRIAL	COMMERCIAL	INSTITUTIONAL	TRANSPORTATION	COLLECTIVE RESIDUALS HANDLING AND MODIFICATION
SIC-117[b] Metal Processing	11,108	Restaurants 2,577	Government offices 3,704	Fuel Sales 183	Recyclable Solid Residuals 265
SIC-120 Chemical Production	3,589	Hotels 910	Research Institutions 3,125	Vehicle Repair 3,740	Mixed Solid Residuals 800
SIC 123 Pulp and Paper Production	2,406	Retail Trade 6,759	University 2,264	Office Services 2,044	Liquid Residuals 370
All Other	22,883	Offices 8,995	Secondary Schools 3,896	All Other 3,928	
		Wholesale Trade 12,356	Cultural 1,076		
			Health Institutions 4,804		
TOTAL	39,986	31,597	19,069	9,985	1,435
% of 102,072	40%	32%	17%	10%	1%

[a] Employment considered only for activities included in REQM analysis. Approximately 3,000 additional persons are employed in agriculture, bringing total employment to about 105,000.

[b] SIC = Yugoslav Standard Industrial Classification.

Source: Tabulated from data in Statistical Letopis SR Slovenije (Ljubljana, Yugoslavia: Zavod za Statistiko, SR Slovenije, 1974).

59

(As previously mentioned, SEPO has adopted the policy of not approving credit for projects with potentially negative AEQ impacts.) All four institutions are based in Ljubljana, the center of the five-commune area. The first and fourth institutions have responsibilities across the total range of environmental problems; the second only with respect to air; the third only with respect to water.

Factors Affecting the Demarcation of the Study Area

Several factors affect the choice of the boundaries of an area for REQM analysis. The following are those which were important in selecting the five-commune area for study.

First, ideally an REQM area should encompass both all of the residuals dischargers and all of the receptors of the changes in AEQ resulting from the discharges of those residuals. About 95 percent of residuals generation within the basin is within the five-commune area. Essentially no gaseous residuals are imported into the basin or the study area, and those generated in the basin are usually trapped within it. Almost all of the solid residuals generated within the five communes are disposed of in existing landfills within the area; no solid residuals are imported into the area. The five-commune area comprises a relatively small portion of the Upper Sava River Basin geographically, as shown in figure 7. The liquid residuals brought into the study area, as a result of activities upstream, are for the most part transported through and out of the study area, along with the liquid residuals discharged within the area.

Figure 7. Upper Sava Basin

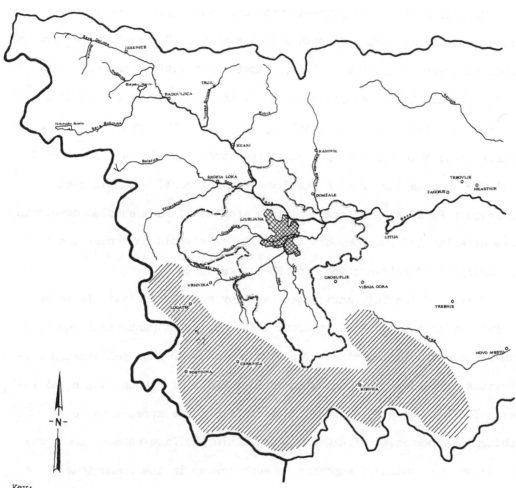

Key:

 Five-Commune Area Boundary

━━━━ Upper Sava River Basin Boundary

/////// Karst Region

XXXXXX Urban Area

Second, an REQM analysis--as with any other type of empirical study--must be accomplished with a finite amount of resources in terms of time, manpower, analytical skills, computer facilities, and existing data. The available resources constrain the combination of the size of the area and the degree of detail applied within the area. Given the resources available for the Ljubljana area study, an adequate level of detail for an area larger than the five communes could not have been achieved. Focusing on a smaller study area would have enabled developing more detailed information, but the areal scale would have been too small for analysis of relevant regional REQM strategies.

Third, if the information generated by an REQM analysis is to be useful, it must be related to areal units for which management decisions can be made and implemented. The five communes are the political jurisdictions in the area, among which various modes of communication and interaction have been developed. In addition, there appears to be a substantial awareness of AEQ problems within the five-commune area, evidenced by the continued emphasis on such issues in the communications media. Among the factors contributing to this awareness are that the area is a political and a university center, thereby attracting much technical and professional expertise, and the location in the five-commune area of a number of nature and environmental preservation organizations.

Fourth, available data on economic activities and on demographic characteristics of the population are compiled by communes. Selecting a study area whose boundaries did not coincide with the statistical bound-

aries would have necessitated expenditure of some of the available resources to modify the available statistics to correspond to the selected
boundaries.

Fifth, the application of REQM analysis to an area requires an institutional base in the area, both to insure that the data used reflect
local conditions and to provide a channel for flow of information generated into the decision-making process. This institutional base was provided by the Urbanisticni Institut SRS, the co-sponsors of the study.
This institute is a workers' self-management enterprise located in
Ljubljana. The Institut has been involved for fifteen years--and still
is involved--with urban planning and development in Slovenia, particularly in the five-commune area. In addition to its role in actual data
collection and analysis for the REQM study, the Institut was responsible
for formal contacts with other cooperating agencies and enterprises.

Finally, because the application of the REQM approach to analysis of
an area was a "first" in Yugoslavia, it was important to select an area
which would have physical and economic characteristics perceived to be
sufficiently similar to other areas in the country so that what was
learned in terms of overall methodology, air quality modeling, and analyzing activities, would be applicable to other areas in the country.

Chapter IV

EVOLUTION OF THE LJUBLJANA REQM ANALYSIS

Introduction

It is axiomatic to state that the degree of sophistication of an
analysis should be "tailored" to the situation: questions being asked;
available analytical resources; and political constraints. The degree of
sophistication should be in tune with the subtleties of the institutional
milieu in which the analysis is made. It is likewise axiomatic to state
that it is essential to plan the analysis itself. That is, the analytical
activities must be scheduled to produce the desired end product, a useful
report, with the available analytical resources.

This chapter describes how these axioms were effectuated in the
Ljubljana study, that is, provides the background on why the component
parts of the analysis were done as they were done. The "why" reflects, to
an important degree, the evolutionary nature of the Ljubljana study. Of
course, any analysis should be evolutionary in the sense that iteration
should take place as data collection and analysis proceed and "learning"
occurs, with modifications being made in study outputs, details of analysis, and allocation of the available analytical resources. But in the
Ljubljana study, the evolution also included changes in the program of
study made possible by significant changes in the available analytical
resources during the course of the study.

The analysis for REQM in the Ljubljana area had to be planned and carried out in a context of substantial uncertainty. This uncertainty had four dimensions. One, available personnel changed over time. At the beginning one full-time U.S. professional, two weeks of U.S. professional consulting services, and an unknown amount of services from Yugoslav enterprises and governmental agencies were available. Subsequently, an additional full-time U.S. professional, approximately twenty-four weeks of professional consulting services, and funds to pay for Yugoslav research assistance became available. In addition, Yugoslav enterprises and governmental agencies contributed manpower, equipment, and laboratory services.

Two, the time span for the study itself changed over time. Originally the study was programmed for one year. Subsequently this was extended to eighteen months, and still later to two years.

Three, it was not known at the start what Yugoslav data relevant to activity models and natural systems models were available, even though a preliminary general description of REQM problems had been prepared prior to the initiation of the formal REQM analysis in July 1973.[16] Only as the study proceeded did it become clear what specific data did and did not exist.

Four, it was not known, originally, the extent to which it would be

---

[16] W.J. Lontz and R.W. Reed, A Report Comparing and Analyzing the Provision of Environmental Protection Services in Ljubljana, Yugoslavia and Grand Rapids, Michigan (Ljubljana, Yugoslavia: Johns Hopkins University/Urbanisticni Institut Urban Planning Project, 1973).

possible to obtain data from existing Yugoslav enterprises and governmental agencies.

In addition to uncertainty, three other factors affected the organization and content of the analysis. They were: (1) an almost complete
lack of Yugoslav data on residuals generation and discharge; (2) the lack
of a single governmental agency responsible for REQM in the area, and
hence the lack of Yugoslav definitions of REQM problems in the area; and
(3) political constraints on data availability and data collection.

The lack of Yugoslav data on residuals generation meant that a substantial portion of the available analytical resources had to be allocated
to producing those data. The second factor had two implications. One,
the lack of a single REQM governmental agency meant that there was no existing locus from which to obtain reactions to project procedures and to
interim outputs as the work proceeded. Consequently, resources had to be
allocated to identify relevant governmental agencies and enterprises for
providing those reactions and to stimulate their responses. Two, the lack
of clear Yugoslav definitions of REQM problems meant that the project had
to define the problems and the corresponding outputs to be produced by the
study. Hopefully, as the study proceeded, those outputs would be perceived
by Yugoslav enterprises and governmental agencies as relevant.

The third factor operated unexpectedly and inconsistently: here today, gone tomorrow. For example, specific data on the meteorological variables used in the air quality model were made available only after some
delay, when the appropriate approvals were obtained. Data on income distribution in Slovenia were not available to the project at all.

The uncertain context and these three factors explicitly conditioned choices made with respect to how the analysis was conducted and how the available analytical resources were allocated and reallocated.

Analysis Conditions

A necessary first step in setting up an REQM study is to define the total set of conditions under which the analysis will be performed. Although it is not usually possible to define accurately, at the outset of a study, all the conditions under which a study will be carried out, a necessary initial step is to define them as completely as possible. For example, decisions must be made on the boundaries of the study area, the methodology to be used for analyzing REQM systems, what variables are assumed fixed, and the time period to be covered. These decisions should be made in relation to the available analytical resources.

The following eight analysis conditions were specified.

1. Residuals generation coefficients to be used were to reflect Yugoslav conditions.

2. REQM strategies would be developed jointly by Yugoslavs and Americans and would reflect technologic, economic, and political feasibility within Yugoslavia.

3. Costs for residuals discharge reduction and modification measures would be developed on the basis of existing Yugoslav costing methods and procurement policies. For example, items stipulated by law to be manufactured in Yugoslavia would be costed as such;

import taxing policies and funding and credit arrangements, as defined by law, would be followed.

4. The study would be limited to the five-commune area which represents the crux of the REQM problem in the Ljubljana region.

5. Where natural systems modeling would not be possible, percent reductions in residuals discharges would be used as environmental quality targets.

6. System variables would be assumed static and deterministic and final demand would be assumed fixed.

7. Locations and levels of activities would be assumed fixed.

8. The analysis would reflect 1975-1976 conditions, using data for the base year, 1972.

The last condition warrants amplification.

The study, based largely on 1972 data, was intended to be relevant to conditions approximately existing through about 1976. It is difficult to evaluate how much beyond that date the results of the study--in contrast to the method--will remain useful. In a rapidly changing society such as that of Yugoslavia, technology, social tastes, and factor prices are changing at a rate which means that estimated residuals generation coefficients and residuals discharge reduction costs will remain valid for only a few years. In addition, levels of activities are changing rapidly.

However, to do an REQM analysis for some time period in the future is a significantly larger task than to do one for present conditions. Even if the economic-demographic data, including the spatial location of activities, were provided as the starting point for an REQM analysis, more or less detailed analyses--depending on the questions to be answered-- would have to be made for each major residuals generator of the effects on its residuals generation of likely changes in technology, product mix, factor prices. Similarly, more or less detailed analyses would have to be made of likely changes in the technologies of materials/energy recovery, by-product production, and residuals modification and disposal.

For the Ljubljana study, an analysis of the future, besides being in-feasible with the analytical resources available, would have been a ques-tionable activity at the time of the study because the plan for the re-gion, "Ljubljana 2000," was incomplete. Also, among the Yugoslavs there was great skepticism about the utility of analyzing hypothesized futures and a corollary hesitation to make estimates of future conditions and to perform analyses relating to them. Consequently, it seemed more useful, at the time, to analyze the more believable present period. The rationale was that an analysis of present conditions would not only be more likely to be useful with respect to current REQM decisions, but that it would develop the capacity to construct activity and natural systems models, demonstrate the application of the REQM framework, and provide a basis for developing activity and natural systems models for a more complex analysis of future conditions at some later date, if that proved to be desirable.

Study Outputs and Method of Analysis

If the outputs of a study are to have any impact on the decision-making process, those outputs must convey positive answers to at least two questions. One, is the information generated by the study "believable" with respect to feasibility of implementation, estimated total costs, and the distribution of costs? Two, is the information in a form which can be understood by decision makers and which can be used as direct inputs to the process for choosing REQM strategies?

Rarely is it possible to investigate the effects of <u>all</u> variables on AEQ and REQM costs. Therefore, selected and specific study outputs must be delineated which are feasible, relevant, and important, given the analysis conditions. The following study outputs were specified:

1. estimate least cost combinations of physical measures to achieve specific levels of environmental quality, where environmental quality is defined by a set of indicators, each of which relates to the ambient concentration of, or to a limitation on the discharge of, a residual; and,

2. demonstrate the effects of particular physical measures on AEQ and REQM costs.

This specification of study outputs was affected by the fact that Yugoslavia has been, and still is, undergoing rapid economic growth, with Slovenia being one of the fastest growing areas of the country. Although there is a clear concern with the increasing deterioration of some aspects of environmental quality which has accompanied that growth, particularly

with respect to certain indicators of air quality, there is also a concern
that the physical measures to improve environmental quality will be too
expensive. Thus, the first study output specified reflects the concern
with finding the least cost strategy to achieve any given level of environmental quality. The second study output defined reflects the view
that, given limited resources in Slovenia to improve environmental quality,
it would be helpful to have information on the effects and costs of particular single REQM measures which are under discussion and/or might be
adopted, such as the provision of space heating for multi-flat residences
from central power plants. Both outputs are consistent with the goals
stated in chapter I.

Specifying levels of environmental quality requires specifying the
residuals to be analyzed. It is seldom possible, and in fact is likely
to be neither desirable nor necessary, to analyze all residuals. Because
not all residuals are likely to be important in a given region, or of equal
importance, some basis must be used to select the residuals to be analyzed.
For the Ljubljana study the following five criteria were used:

1. that the residual be relatively important with respect to other
   residuals, in terms of residuals management costs and/or adverse
   effects on environmental quality;

2. that for the important residuals there be either available data
   on generation or that institutional arrangements be available for
   collecting raw data on generation;

3. that the residual be a <u>generally</u> accepted indicator of environ-

    mental quality;

4. that there be "known" physical measures for reduction of the dis-

    charge of the residual to, or disposal in, the environment; and

5. that the analysis of the number of residuals selected be feasible

    within the analytical resources available.

Table 6 shows the environmental quality indicators and the target

levels selected for analysis. The latter represent Yugoslav environ-

mental quality standards where possible. Because no natural systems

models were developed for residuals other than sulfur dioxide and par-

ticulates, environmental quality targets for the other residuals were

specified in terms of percent reduction of residuals discharges. At the

time of the study there were no BOD_5 and TSS effluent standards and nei-

ther ambient standards nor effluent standards for CO, HC, and NO_x in

Yugoslavia. The environmental quality targets shown in table 6 represent

the base set of environmental quality targets. Other sets of environmental

quality targets were subsequently specified in order to illustrate various

levels of environmental quality which could be achieved at different costs.

Given the available analytical resources, it was clear from the start

of the study that only a relatively few combinations of physical measures

could be investigated. That fact, and having defined five of the seven

environmental quality targets in terms of percent reduction in discharge,

lead to the selection of the search method of analysis. This meant that

the available analytical resources would be concentrated on identifying,

Table 6. Environmental Quality Indicators Selected for Analysis and Base
Level of Environmental Quality

Environmental Quality Indicator	Abbreviation	Targets for base level of E.Q.: degree of discharge reduction; ambient concentration; quality of landfill operation
Liquid		
Five-Day Biochemical Oxygen Demand	BOD_5	80%
Total Suspended Solids	TSS	80%
Gaseous		
Sulfur Dioxide	SO_2	150 $\mu g/m^3$ [a,b]
Total Suspended Particulates[c]	TSP	150 $\mu g/m^3$ [a,b]
Carbon Monoxide	CO	50%
Hydrocarbons	HC	50%
Nitrogen Oxides	NO_x	10%
Solid		
Mixed Solid Residuals (primary and secondary)[d]	MSR	Good quality sanitary landfill

[a]1972 Yugoslav standard; μg= micrograms.

[b]Highest seven-day average during a four-month period

[c]Total suspended particulates refer to the measurement of the ambient concentration of suspended particles of various sizes from various sources. The term particulates refers to the discharge of particulate matter from exhaust stacks, some portion of which comprises the measured total suspended particulate concentration.

[d]Secondary residuals are those generated in the process of modifying primary residuals, e.g., those originally generated.

and developing costs for, physical measures to reduce residuals discharges, with no resources having to be allocated to setting up and running a programming algorithm.

Allocating Analytical Resources

Whatever analytical resources are available in a given study context must be allocated. Figure 8 shows that an explicit allocation occurred at three points during the Ljubljana study. Each time the overall objective was to allocate available resources such that one or more combinations of physical measures relating to the base set of environmental quality targets would be completed and a final report prepared, while maintaining reasonable consistency in detail and accuracy among the elements of the REQM system. At the same time, the reallocations reflected what had been learned up to the respective points in time, about the relative importance of different residuals, different environmental quality problems, and physical measures considered relevant by the Yugoslavs. Within the overall objective the decision was made to give more emphasis to primary data collection on residuals generation than to other components of the analysis. This was done to insure that the completed study would at least provide some empirical data based on Yugoslav conditions, so that there would be a credible basis for decisions on REQM in the five-commune area.

Based on the above considerations, priorities for the analysis were established. The first priority was the formulation of reasonable, yet

Figure 8.   Summary of Sequence of Activities of Ljubljana REQM Analysis

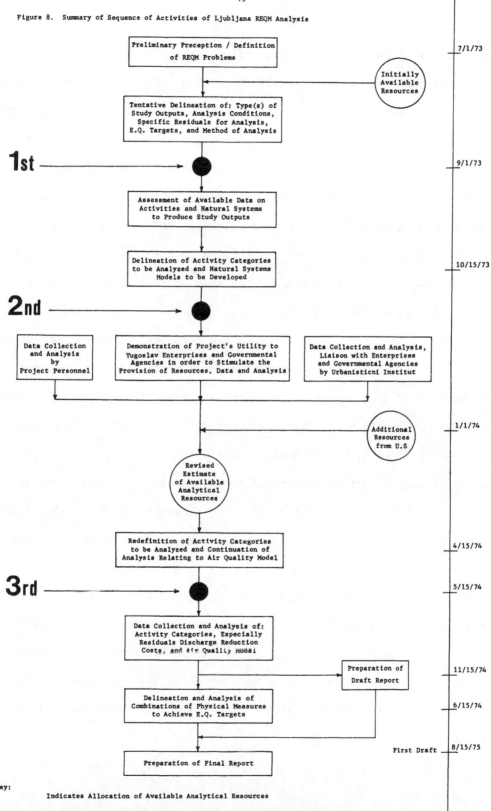

Key:

Indicates Allocation of Available Analytical Resources

simple, activity models which would reflect present conditions and shed
light on factors affecting residuals generation. This in turn meant that
at least some sampling of residuals generation by various activities would
have to be done, given the existing lack of Yugoslav data. The second
priority was the identification of and estimation of costs of physical
measures to reduce residuals discharges, because it was not known at the
outset to what degree natural systems modeling was either possible or useful. The development of natural systems models then became the third priority. Such models would be constructed only where it was reasonable,
simple, and inexpensive to do so.

However, the priorities changed as the study progressed. Increased
knowledge of what data were, or could be made, available and increased
understanding of the concern in the Ljubljana area with air quality, lead
to a shift of available analytical resources to more extensive development
of, and analysis with, an air quality model.

Specifying the nature or type of study outputs is not the same as
specifying the number of each type of study output to be produced, in this
case the number of combinations of physical measures to be analyzed and
the number of sets of environmental quality targets to be investigated.
In the initial period of the study, the known available resources--listed
previously--appeared sufficient to delineate relatively few combinations
of physical measures for an highly aggregated set of activity categories
in relation to only the base set of environmental quality targets. For
example, at that time differentiating among types of residences was not
contemplated.

During the second period of the study, the available analytical resources were increased in two ways. First, relationships had been established with Yugoslav agencies and their interests in the study stimulated.[17]
This lead to the provision by those agencies of data, resources for ac-
quiring data, and review capacity for the study. Second, additional U.S.
professional manpower and funds were made available. These increased resources: enabled greater disaggregation of the activity categories to be
analyzed; expanded the capacity to obtain primary data on residuals generation; expanded the range of physical measures which could be investigated for each activity category; and enabled the analysis of more combinations of physical measures in relation to more sets of environmental
quality targets, five sets as compared to the one set initially specified.

One part of the study for which resource requirements were substantially underestimated was the preparation of the final report. Several
factors contributed to this. One was the international context of the
study, and hence the difficulty in making final checks on points involving
information available only in Yugoslavia. Another was the fact that part
of the team preparing the final report was in Yugslavia, part in the U.S.
The final factor was the lack of knowledge of the final disposition of the
report, that is, whether or not it would be formally published for distribution outside of, as well as within, Yugoslavia.

---

[17]An important means of stimulating interest was the production of
interim reports in both English and Slovene.

In sum, the allocation of analytical resources is an essential, a critical, and a difficult task in undertaking an REQM analysis. One cannot necessarily perceive a priori: (1) what problems may be encountered in making the analysis; (2) the relative importance of different components of the analysis; (3) the relative importance of the different residuals generating activities; and (4) the amount of time the final report will take to complete for its intended audience. The normal allocation problem was compounded in the Ljubljana study by the uncertainty over, and the changing magnitude of, the analytical resources available.

CHAPTER V

ACTIVITY MODELING: ESTIMATING RESIDUALS GENERATION

In chapters V through VIII, the procedures and methods used to develop
the detailed information for REQM decision making in the five-commune area
and the information resulting therefrom are presented. This chapter discusses the separation of activities in the five-commune area into activity
categories and subcategories and the estimation of residuals generation
coefficients and of residuals generation.

The terms "activity model" and activity modeling are used in this report in a broader sense than these terms are used in the economic literature. Herein activity modeling refers to the procedure by which estimates
are made of residuals generation, residuals discharge, and residuals discharge reduction costs for each of the activities in the region. As
indicated in chapter II, activity models can vary from simple to very
complex.[18] A simple model represents residuals generation by a given
activity simply by a set of fixed coefficients, e.g., residuals generated
per unit of activity, under some set of factor input prices, production or
use technology, and governmental policies, all of which are implicit, or
are occasionally explicitly stated. Given the types and magnitudes of residuals generated, residuals modification by the activity is characterized
by a single physical measure "added on" to the activity and applied at a

_____

[18] Examples of activity modeling are discussed in, B.T. Bower,
"Studies of Residuals Management in Industry," in E.S. Mills, ed.,
Economic Analysis of Environmental Problems (New York: National Bureau
of Economic Research, 1975).

single level, e.g., 80 percent removal of BOD_5 by sedimentation plus activated sludge, with the associated costs. An example of a much more complicated activity model is one which represents residuals generation and discharge in the form of a linear program. The programming formulation could include alternative raw material inputs, alternative production technologies, alternative product output specifications, alternative types and levels of application of each type of residuals modification-- all with their associated costs--with various combinations of factor input prices and limits or charges on residuals discharges.[19]

Separating Activities into Activity Categories

In delineating the activity categories, all activities in the five-commune area were first classified by economic function, based on the available data. Then, given the residuals of interest as determined by the environmental quality indicators selected, activity categories were established on the basis of four factors. The first was the obvious one of which activities generate the residuals which affect the chosen environmental quality indicators. The second was the relative importance of a particular residuals generating activity with respect to other generators of the same residual, e.g., what proportion of total generation of each residual is attributable to the given activity. The third was whether or not an identified activity category can be affected by a

[19]For two excellent examples of the programming approach to activity modeling see C.S. Russell, Residuals Management in Industry: A Case Study of Petroleum Refining (Baltimore: Johns Hopkins University Press, 1973) and C.S. Russell and W.J. Vaughan, Steel Production: Processes, Products, and Residuals (Baltimore: Johns Hopkins University Press, 1976).

feasible REQM strategy. The fourth consisted of the available analytical resources.

Seven activity categories were delineated: industrial; commercial; residential; institutional; transportation; power plant; and collective residuals handling and modification activities, such as municipal incinerators and municipal sewage treatment plants. No agricultural activity category was included because the contribution of agricultural operations to total residuals generation appeared to be small relative to generation by other activities in the five-commune area. These seven categories were subdivided into thirty subcategories, which are listed in table 7. The residential and industrial activity categories are used to illustrate considerations in delineating subcategories.

Residential activities were separated into two subcategories: (1) multi-flat residences are structures comprised of more than two family dwelling units; and (2) single-flat residences are structures comprised of one or two family dwelling units. Review of the census data on housing indicated that there would be significant differences in residuals generation between these two types of residences because of differences in space heating technology and types of fuels used, plumbing, building designs, internal space per dwelling unit, ownership, and socioeconomic status. For example, 70 percent of all flats in multi-flat residences have bath facilities, compared to 59 percent of the single-flat residences. Almost all single-flat residences have yard and garden areas which are watered; multi-flat residences usually have none. About three-fourths of the

Table 7. Activity Categories and Subcategories for REQM
Analysis of Five-Commune Area

INDUSTRIAL ACTIVITIES

Metal processing
Electrical products
Chemical
Building materials
Pulp and paper
Paper products
Textile
Food products
Graphics
All other

RESIDENTIAL ACTIVITIES

Multi-flat
Single-flat

COMMERCIAL ACTIVITIES

Restaurants
Hotels and motels
Retail stores
Offices
Wholesale stores and
warehouses

INSTITUTIONAL ACTIVITIES

Government
Research
Hospitals
Cultural institutions[a]
University facilities
Schools other than university

TRANSPORTATION ACTIVITIES

Vehicular travel[b]
Services[c]

POWER PLANT ACTIVITIES

Moste plant
Siska plant

COLLECTIVE RESIDUALS HANDLING
AND MODIFICATION ACTIVITIES

Municipal solid residuals
collection agency
Municipal sewerage agency
Salvaging operations

a Refers to cinemas, playhouses, art galleries, museums, etc.
b Vehicular travel is further subdivided into seven road types in urban
and nonurban areas.
c Refers to service stations, garages, repair shops.

multi-flat residences have on-site, large capacity energy conversion units installed and operated by housing maintenance companies.[20] The remaining multi-flat residences are heated from central power plants. In contrast, virtually all single-flat residences have very small capacity energy conversion units of various types of technology to provide space heating and hot water for household uses. Very few single-flat residences obtain heat from central power plants.

Mixed solid residuals generation also differs between single-flat residences and multi-flat residences. Yard wastes and ash from solid fuel energy conversion processes can be significant components of residential mixed solid residuals generation. Depending upon the time of year, yard wastes and/or ash account for 50-60 percent of mixed solid residuals generation in single-flat residences. Because energy conversion technologies and the types and quantities of fuels used are different for single-flat residences and multi-flat residences, the corresponding ash components of mixed solid residuals generation differ.

The foregoing factors result in sufficient differences in residuals generation between the two types of residences to warrant separating the two types of residential activity, provided that REQM strategies and/or REQM costs can be differentiated between the two subcategories. Although additional data collection is required because of this disaggregation,

---

[20]Stanovajski Podjetje Fond and Dom are the two principal housing maintenance companies in the five-commune area.

more accurate estimation of residuals generation and delineation of physical measures for reducing residuals discharges and their associated costs is possible than if all residences were placed in one category.[21]

Based on the Yugoslav standard industrial classification (SIC), industrial activities were subdivided primarily in relation to production processes and product mixes, and to a limited extent by the type(s) of raw material(s) used. These three variables, in addition to prices of factor inputs, are the major ones affecting residuals generation in industrial activities. For the Ljubljana study, ten industrial subcategories were delineated: nine SIC categories and one for "all other industries." Subcategories for the other five activity categories were delineated in an analogous manner.

Estimating Residuals Generation

Introduction

For each activity subcategory some parameter must be selected which indicates the level of activity, e.g., tons of product output, kilocalories of energy generated, barrels of crude petroleum processed, number of employees, number of inhabitants, vehicle kilometers traveled. The number of employees is usually readily available information and is frequently used in regional planning to specify present, and in projecting future, levels of different activities. For industrial activities it is

----

[21]Information based on the two subcategories is also useful for routing and scheduling mixed solid residuals collection.

desirable to use physical units of input or output, because of the direct relationship between these physical units and residuals generation. Residuals generation coefficients are then expressed in terms of residuals generated per the relevant unit for each subcategory. For example: for an industrial activity mixed solid residuals generation is expressed in kilograms per metric ton of product output; BOD_5 generation in residences is expressed in grams per day per inhabitant; HC generation by automobiles is expressed in grams per vehicle kilometer traveled on a particular type of road. The unit used for each activity subcategory and the 1972 level of activity for each subcategory are shown in table 8.

That all of the nonproduct outputs of materials and energy from an activity are not necessarily residuals should be emphasized. Some degree of materials and/or energy recovery may occur, depending on the value of the recovered materials and energy in relation to alternative sources of equivalent materials and energy with use of the recovered flows as inputs into either the same activity or another activity. For example, whenever solid nonproduct outputs in the five-commune area were recovered and sold, these quantities were not included in the quantities of solid residuals estimated. Explicit consideration of recovery was not necessary for liquid and gaseous residuals, because either no recovery took place or recovery was internal to the activity.

In Yugoslavia at the time of the study, there was little or no information, published or unpublished, on residuals generation or discharge. Consequently, residuals generation coefficients for each activity

Table 8. Activity Categories and Subcategories and 1972 Levels of Activity, Five-Commune Area

Activity Category and Subcategory	Level of Activity, 1972	Unit
INDUSTRIAL ACTIVITIES		
Metal processing	94,382	Ton of P.O.
Electrical products	19,583	"
Chemical	99,940	"
Building materials	891,703	"
Pulp and paper	59,312	"
Paper products	27,171	"
Textile	11,541	"
Food products	173,049	"
Graphics	26,579	"
All other	58,743	"
RESIDENTIAL ACTIVITIES		
Multi-flat	165,756	Inhabitants
Single-flat	84,230	"
COMMERCIAL ACTIVITIES		
Restaurants	2,577	Employees
Hotels and motels	910	"
Retail stores	6,759	"
Offices	8,995	"
Wholesale stores and warehouses	12,356	"
INSTITUTIONAL ACTIVITIES		
Government	3,704	Employees
Research	3,125	"
Hospitals	4,804	"
	74,563	Patients[a]
Cultural institutions[b]	1,076	Employees
University facilities	2,264	"
Schools other than university	48,315	Students
TRANSPORTATION ACTIVITIES		
Vehicular travel[c]	985×10^6	Vkt: Gas
	110×10^6	Vkt: Diesel
Services[d]	9,895	Employees
POWER PLANT ACTIVITIES		
Moste plant	1.427×10^{12}	Calories
Siska plant	0.057×10^{12}	"
COLLECTIVE RESIDUALS HANDLING AND MODIFICATION ACTIVITIES		
Municipal solid residuals collection agency	800	Employees
Municipal sewerage agency	370	"
Salvaging operations	265	"

Abbreviations: P.O., product output; Vkt, vehicle kilometers traveled.

a Average stay in hospital, 17.8 days.
b Refers to cinemas, playhouses, art galleries, museums, etc.
c Vehicular travel is further subdivided into seven road types in urban and nonurban areas.
d Refers to service stations, garages, repair shops.

subcategory were developed by one or the other of two approaches: (1) adopting and/or modifying coefficients found in the literature; (2) deriving coefficients from empirical data collection and analysis. For each activity subcategory the coefficient estimated by either approach represents the mean value for the subcategory; some individual operations in each activity subcategory have higher, some lower, generation coefficients.

In applying the first approach, much caution had to be used in adopting coefficients from non-Yugoslavian sources which reflected different cultural and economic conditions. For example, a coefficient for residential BOD_5 generation developed in the Federal Republic of Germany is 54 g/person/day, which is defined as one population equivalent (P.E.). However, sampling in the five-commune area produced a coefficient of 69 g/person/day, a substantial difference. Thus, residuals generation coefficients from literature from outside Yugoslavia were used only where they seemed to be reasonably representative of specific Yugoslav conditions. U.S. coefficients were adopted for: BOD_5 and TSS for the food products and pulp and paper industries, for which industries production processes matched their U.S. counterparts; SO_2 and particulates for all activities for which energy conversion technology could be matched with Yugoslav technologies; and CO, HC, and NO_x based on types of fuel combusted in all activities. For vehicular travel, West German coefficients for CO, HC, and NO_x discharges were used, based on vehicle types similar to those in the five-commune area.

For activity subcategories for which factors affecting residuals generation outside of Yugoslavia differed from Yugoslav conditions, empirical data collection was undertaken. Given the limited resources, only limited sampling was possible, even with the assistance of local agencies. For example, excluding wastewater discharges from power plants, wastewater discharges from industrial and residential activities represented 52 percent and 32 percent, respectively, of regional wastewater flow. All other activities represented only 16 percent. Therefore, for estimating WW, BOD_5, and TSS generation coefficients, sampling was concentrated on the industrial subcategories other than food products and pulp and paper, and on the residential subcategories. Because no data were available with respect to MSR generation coefficients, sampling of all activity subcategories was undertaken.

In all cases sampling was done with respect to the individual activity as a whole, not with respect to the unit processes and unit operations within the activity. Estimating residuals generation coefficients without explicit analysis of factor input prices and analysis of unit processes and operations within an activity limits the range of residuals discharge reduction measures which can be analyzed and precludes analysis of the impacts of changes in factor inputs. Therefore, for illustrative purposes, two activities were selected for application of a more detailed level of activity analysis, by means of explicit materials and energy balances. At least that level of detail would be

relevant in any area where only one or a few activities produce
most of the residuals generated and would incur most of the REQM costs
to achieve desired AEQ standards.

The first activity was an industrial plant, the "Union" brewery.
This is one of the largest industrial enterprises in Ljubljana and generates large quantities of wastewater and various other residuals. Data
collection at, and analysis of, the brewery illustrated the complexity of
modeling an activity with multiple processes and operations, in this case
the production of malt, yeast, and beer and the production of steam for
process use. Figure 9 shows the materials balance for the production of
beer in 1972. Table 9 shows residuals generation for the entire brewery
in 1972.

The second activity was the coal-fired Moste power plant of the
Toplarna enterprise. The materials balance was made for the plant as a
whole, and is shown in figure 10. Analyzing activities even at this
level of detail, e.g., for the activity in the aggregate, requires a
large amount of analytical resources if it is to be applied to numerous
activities in a regional REQM analysis.

Estimating Mixed Solid Residuals (MSR) Generation Coefficients

MSR generation coefficients were estimated for each activity by
sampling actual MSR generation with the cooperation of the municipal
solid residuals collection agency in Ljubljana. A number of sites were
selected for each activity subcategory, at which solid residuals over a
period of time were collected and weighed. The related level of activity

Figure 9. Materials Balance For Beer Production in Union Brewery, 1972

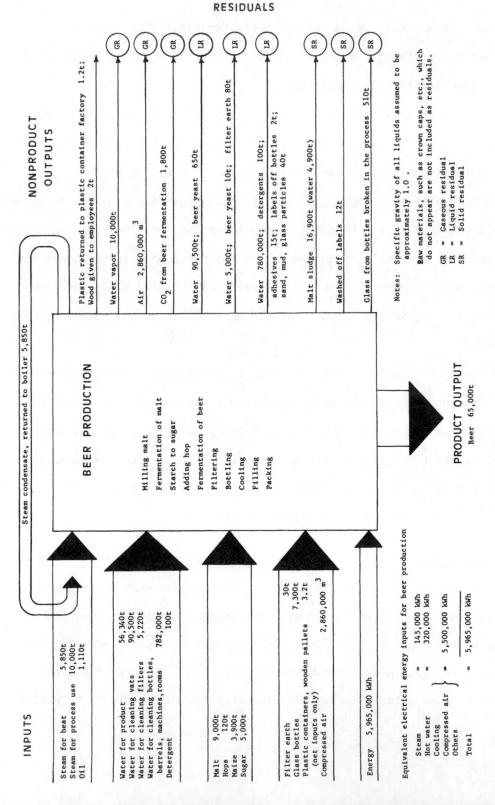

Source: Modified from : B.T.Bower, D.J. Basta, Residuals Environmental Quality Management: Applying the Concept (Baltimore, Johns Hopkins University Center for Metropolitan Planning and Research, October 1973).

Table 9. Residuals Generation in the Production of 1,000 Liters of Beer,[a]
"Union" Brewery, 1972

(Product Output Mix: 2% "Bock" (16%); 3% "Lezak" (14%); 95% "Triglav" (12%)[b])

RESIDUALS GENERATED	10^{-3} METRIC TONS
Gaseous:	
- CO_2	29.70
- Air	4.40 (cubic meters)
Liquid:	
- H_2O, earth, rye fibers	927.00
- H_2O, beer yeast, filter earth	77.50
- H_2O, beer yeast	140.10
- H_2O, detergents	1200.00
Solid:	
- Rye sprouts	4.76
- Glass	7.85
- Plastics	0.02
- Wood	0.03
- Mixed solids	
adhesives	
labels	
glass particles	
sand	
other inerts	0.60
- Malt sludge	26.10

a One metric ton of beer.
b Percentages in parentheses are alcoholic contents.

Source: B.T. Bower and D.J. Basta, Residuals Environmental Quality
Management: Applying the Concept (Baltimore: Johns Hopkins
University Center for Metropolitan Planning and Research,
October 1973).

Figure 10. Overall Materials Balance For Moste Coal Fired Power Plant, 1972 [a]

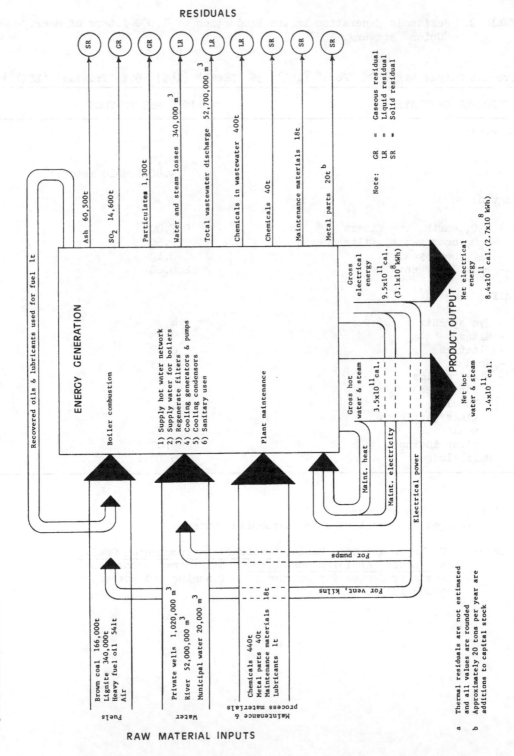

Note: GR = Gaseous residual
       LR = Liquid residual
       SR = Solid residual

a   Thermal residuals are not estimated
    and all values are rounded

b   Approximately 20 tons per year are
    additions to capital stock

was determined for the corresponding time period. The coefficient was calculated in terms of MSR generated per unit of activity. Sampling activities were carried out over approximately a four month period for twenty-one of thirty activity subcategories. For eight activity subcategories, coefficients were developed based on engineering estimates. For the vehicular travel activity subcategory MSR generation was negligible. Details of the estimation methods for industrial and residential activity categories are described below.

Estimating Industrial MSR Generation Coefficients. A preliminary investigation revealed that significant quantities of industrially generated nonproduct solids, e.g., iron, steel, rubber, were sold to salvaging firms[22] in the Ljubljana area, which in turn sold them as inputs to other activities. In order to estimate MSR generation coefficients under present conditions, the quantities of both "salable" and "nonsalable" components of the nonproduct solids generated had to be determined.

The complete records of solid materials purchased by the salvage firms for 1972 were analyzed. These records indicated the quantities and types of solid materials purchased from each industrial plant or enterprise. Each of these activities was classified into one of the ten industrial subcategories used. The total salable nonproduct solids generated by each subcategory could then be estimated.

---

[22]There are three major salvaging firms in the area: Dinos, Surovinos and Unija. Unija collects primarily used paper products, whereas the others collect all types of marketable materials.

The quantity of nonsalable nonproduct solids, e.g., the MSR generated by each industrial subcategory, was estimated by analyzing available data from the collection schedules of the municipal MSR collection agency. These schedules indicated the total volume of MSR collected in 1974 from each industrial plant, including those already identified as generating salable nonproduct solids. A program of sampling MSR collected from several plants in each industrial subcategory was carried out, to enable estimating the mean density of MSR generated in each subcategory. The mean density was applied to the total volume collected from all the plants in the subcategory for the year, to estimate the weight generated in the year. Total weights of salable and nonsalable nonproduct solids generated in each subcategory were then related to the total tons of product output to obtain coefficients of total nonproduct solids, and MSR generated, per ton of product output.[23] Table 10 shows the percentage of the total volume of MSR generated in the sampling period by each subcategory which was sampled and the percentage of the total volume of MSR generated in the sampling period by each subcategory.

---

[23]1972 production levels (annual tons of product output) were used with 1973 salvage firm data and 1974 volumetric data to derive the coefficients. This mixing of information from different years was necessary because it was not possible to obtain the needed data for the same year. As far as could be ascertained, the three variables were relatively constant over the period.

Table 10. Percentage of MSR Sampled by Industrial Subcategory

Industrial activity subcategory	Percentage of total MSR volume generated by subcategory which was sampled	Percentage of total volume of industrial MSR generated in sampling period
Metal processing	5	15.6
Electrical products	11	7.0
Chemical	5	11.3
Building materials	Included in All other	Included in All other
Pulp and paper } Paper products }	23	17.6
Textile	20	2.4
Food products	44	19.2
Graphics	45	2.4
All other	8	24.5
		100.0

Estimating Residential MSR Generation Coefficients. Seasonal variation in MSR generation is particularly important in residential activities,
both single-flat residences and multi-flat residences. Because it was not
possible to sample during each season, results of sampling in one season
were adjusted to account for the seasonal variation in MSR generation, as
described below. Seasonal variation is a result of two major factors.
First, because single-flat residences have yard space, MSR generation
during the spring, summer, and fall seasons is affected by the generation
of yard wastes. Multi-flat residences have little or no yard space, and
hence little or no yard waste is generated. Second, space heating varies
substantially by time of year, which results in seasonal differences in
ash generation. Separate sampling of both types of residences was performed to develop MSR coefficients.

For multi-flat residences it was possible to choose sampling locations which were minimally affected by seasonal factors. Locations were chosen which: (1) could be sampled by dumpster pickup to facilitate collection; (2) included at least 100 inhabitants; and (3) were heated by an essentially ash-free technology. In five buildings at four locations which met these criteria, MSR generation was sampled during a two-week period in the spring. The total weight of MSR collected during the period from the five buildings divided by the total number of inhabitants yielded the MSR generation coefficient, exclusive of ash. The ash component of MSR generation in multi-flat residences was estimated as described below.

For single-flat residences, a "typical" neighborhood of approximately 1,500 people in the five-commune area was selected for sampling. Selection was based on four criteria: (1) homogeneity, e.g., almost all single-flat residences; (2) known number of inhabitants; (3) feasibility of a single truck collection route; and (4) feasibility of identifying structures which were not single-flat residences and omitting them without inordinately disrupting regular MSR collection. Sampling was done in the spring and coincided with regular collection schedules during a two-week period. Each day the total weight of MSR collected from the single-flat residences, including ash and yard wastes, was tabulated. The total weight of MSR collected divided by the known number of inhabitants yielded the MSR generation coefficient.

The composition of the MSR collected in the sampling of single-flat residences and multi-flat residences was determined in cooperation with the municipal MSR collection agency. The results are shown in table 11. The

Table 11. Composition of MSR Generation in Residences, 1972[a]

Components of Mixed Solid Residuals	Percent of Total MSR Generated in:	
	Single-Flat Residences	Multi-Flat Residences with Space Heated from Power Plant
Metal	4	5
Glass	3	9
General food wastes	10	35
Paper	14	31
Plastics	4	7
Textiles	3	7
Ash[b]	8	None[d]
Yard wastes (Total)	54	None
Stones	9	
Earth and other[c]	45	
Other[c]	In yard wastes	7
TOTAL	100	100

a Based on sampling in the spring.
b Ash content calculated as described in text.
c Includes unidentifiable residuals less than
   30 millimeters in size.
d No ash is generated on-site for a building
   centrally heated by an external source.

amount of paper generated per inhabitant is primarily a function of the
amount of packaging.[24] Each item purchased, such as a bottle of wine or
a bottle of hair tonic or a pencil, is individually wrapped and then put
into a paper bag. The relatively large component of general food wastes
is a result of the small amount of processed--canned, frozen, dried--food
used. Almost all vegetables and fruits are sold fresh. Most of the meat
and poultry is unpackaged.

The percentage of MSR generated attributable to yard wastes found in
sampling the collections from single-flat residences in the spring was
used to estimate the seasonal pattern of generation of yard wastes in
single-flat residences. Experience of the municipal MSR collection agency
indicated that the largest generation of yard wastes is during the spring/
fall period; the smallest during the winter season. Winter season generation was estimated to be about 15 percent of the spring/fall generation;
summer season generation was estimated to be about 80 percent of spring/
fall generation.

To estimate the seasonal ash component of MSR generation for each
residential category, all ash residuals generated from fuel combustion for
residential space heating were estimated to be disposed of through MSR
collection. Ash from all other sources in residences was estimated to be
negligible. Based on fuel composition and consumption data for all of

---

[24]Newspapers are small in size and quantities of newspapers and
periodicals purchased per inhabitant are much less than in the U.S.

1972, seasonal ash generation coefficients were estimated for single-flat residences and multi-flat residences using the following formula:

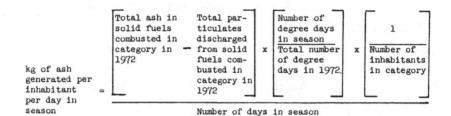

$$
\begin{array}{c}
\text{kg of ash generated per inhabitant per day in season}
\end{array}
=
\frac{
\left[\begin{array}{l}\text{Total ash in solid fuels combusted in category in 1972}\end{array}-\begin{array}{l}\text{Total particulates discharged from solid fuels combusted in category in 1972}\end{array}\right] \times \left[\begin{array}{l}\text{Number of degree days in season}\\ \hline \text{Total number of degree days in 1972.}\end{array}\right] \times \left[\begin{array}{c}1\\ \hline \text{Number of inhabitants in category}\end{array}\right]
}{\text{Number of days in season}}
$$

The formula was applied to each residential category for each season. The year was divided into three periods: (1) winter season, November, December, January, and February; (2) spring/fall season, March, April, September, October; and (3) summer season, May, June, July, August. A degree day is defined as a one degree difference from a reference temperature for a given day. For the Ljubljana area the reference temperature used is $12^{\circ}C$.

Table 12 shows the estimated mean seasonal and mean annual MSR generation coefficients for single-flat and multi-flat residences for 1972, based on the above analyses. The "general" component of MSR for single-flat residences is about 1.3 times that for multi-flat residences. One factor contributing to this difference consists of maintenance and recreational activities characteristic of the former but not the latter. A factor which would reduce the difference is that some of the paper residuals generated in single-flat residences are burned in the space heating units in such residences, an option not available in most multi-flat residences. There may be other factors which also contribute in both directions to the difference.

Table 12. Estimated Mean Seasonal MSR Generation Coefficients for
Residences, 1972
(All values in kg per inhabitant per day, as collected.)

	Winter	Spring/Fall	Summer
Single-Flat Residences			
Yard wastes[a]	0.06	0.43	0.39
Ash	0.26	0.07	< 0.01
General[b]	0.31	0.31	0.31
TOTAL	0.63	0.81	0.70
Multi-Flat Residences			
Ash	0.20	0.05	< 0.01
General[b]	0.24	0.24	0.24
TOTAL	0.44	0.29	0.25

a Yard wastes include stones, earth, grass, trimmings, etc.
b General includes metal, glass, general food wastes, paper, plastics.

Estimating Liquid Residuals Generation Coefficients

Because essentially no liquid residuals modification (wastewater
treatment) facilities are in place at individual activities or enterprises,
liquid residuals generation is equal to liquid residuals discharge.
Therefore liquid residuals generation coefficients were developed for all
activities by estimating discharge coefficients. Wastewater (WW) discharge
coefficients were estimated for all activities, and BOD_5 and TSS coefficients were estimated for all except the institutional and commercial
activity categories. These two categories were assigned the same unit
BOD_5 and TSS coefficients used for the residential category, even though
it is considered likely that this overstates the unit generation for the
institutional and commercial activity categories.

First, estimates of or data on wastewater discharges were obtained in order to identify the larger dischargers, industrial and other. On the basis of these data, the limited resources were allocated to sampling discharges to obtain BOD_5 and TSS data. For plants in the industrial category, water intake data for <u>each</u> plant for 1972 were obtained from industrial plant monitoring records.[25] Wastewater discharge for each plant was then computed by making engineering estimates of consumptive use of water--water which is evaporated, transpired, and/or incorporated in a product, and not discharged--as a percent of total water intake for each plant.

With respect to the residential, commercial, institutional, and transportation services categories, water intake bills were obtained for 1972 for a representative sample of <u>individual</u> activities in each subcategory. Wastewater discharges were then estimated by assuming consumptive use was 5 percent of water intake. For the collective residuals handling and modification activities, engineering designs provided the bases for estimating wastewater discharges. For the power plant category, materials balances were made for the two power plants to obtain wastewater discharges.

Table 13 shows the estimated percentages of total wastewater discharge in the five-commune area by activity category and subcategory, excluding discharges from power plants. The table indicates clearly the

---

[25]These records were made available by Zavod za Vodno Gospodarstvo (Office for Water Usage), Ljubljana.

Table 13. Estimated Percentage of Five-Commune Area Wastewater Discharge,[a] 1972, by Activity Category and Subcategory

Activity Category	Percent of Five-Commune Area Wastewater Discharge
INDUSTRIAL	
Metal processing	4
Electrical products	1
Chemical	6
Building materials	Included in All other
Pulp and paper	26
Paper products[b]	N
Textile	2
Food products	10
Graphics	1
All other	2
TOTAL	52
RESIDENTIAL	
Multi-flat	23
Single-flat	9
TOTAL	32
All other	16
TOTAL	100

a Excluding power plant discharges.
b Paper products activities result in negligible wastewater discharges.

major wastewater dischargers in the area, and the larger discharges in
the industrial activity category.

Estimating Industrial BOD_5 and TSS Generation Coefficients. Two or
three "representative" plants were selected from each of the metal processing, electrical products, chemical, textile, graphics, and all other
(including building materials) subcategories for sampling[26] to determine
BOD_5 and TSS concentrations. The plants selected were based on engineering experience with respect to process water use and the associated
residuals loading. (No plants in the pulp and paper and food products
subcategories were sampled because, based on the available data, U.S.
residuals generation coefficients were estimated to be valid for these
subcategories.)

Each industrial plant selected for sampling was surveyed, both to
find all discharge points and to prepare a piping diagram in order to
locate sampling points in relation to unit processes and unit operations.
In some cases, the physical layout of the selected plant made sampling
impossible. In such cases an alternative plant was selected. Four or
five grab samples were taken during a day in relation to each unit process
in the selected plant[27] from which corresponding estimates of BOD_5 and

---

[26]Sampling of both industrial and residential wastewater discharges
was performed jointly with the municipal sewer agency, Kanalazacije, which
also performed all laboratory tests for BOD_5 and TSS.

[27]Given the variation from day-to-day in wastewater and residuals
discharges from industrial plants under normal operations, considerably
more extensive sampling would be necessary to obtain accurate estimates of
mean discharges and variations around the mean. Resources were simply not
available to undertake more extensive sampling. As limited as the sampling
was, it generated the only data available on BOD_5 and TSS discharges.

TSS concentrations in the wastewater discharge from the plant were de-termined. Table 14 shows the proportion of wastewater discharge sampled in each industrial subcategory in which sampling was done.

Mean BOD_5 and TSS concentrations were then computed for each of the subcategories in table 14, using the total number of concentrations de-termined in the sampling. The alternative would have been to compute the mean concentration for each plant sampled, and then the mean of the means, to represent the mean concentration for the subcategory. However, given: (1) that only two or three plants could be sampled in each subcategory; (2) the substantial variation over time in wastewater discharge and con-centration for a single plant; and (3) the substantial variation in waste-water discharge and discharge per ton of product among plants in a sub-category--as shown in table 15 for the textile subcategory--it was con-sidered that the procedure adopted reflected more adequately the varia-tions in concentration across a given subcategory and hence the mean concentrations for the subcategory.

The mean BOD_5 and TSS concentrations for the subcategory were then applied to the mean wastewater discharge per ton for each plant in the subcategory, to obtain the estimated mean BOD_5 and TSS generation co-efficients in kilograms per ton for each plant. The means of these were then computed to derive the two generation coefficients for the sub-category.

Table 14.   Percentages of Industrial Subcategory Wastewater Discharge
            Sampled

INDUSTRIAL SUBCATEGORY	Percentage of waste-water from subcategory sampled	Number of plants sampled in subcategory	Total number of plants in subcategory
Metal processing	1.0	3	23
Electrical products	18.0	3	11
Chemical	2.5	2	12
Building materials	Included in All other	Included in All other	Included in All other
Pulp and paper[a]	0.0	0	2
Paper products[a]	0.0	0	2
Textile	45.0	3	11
Food products[a]	0.0	0	9
Graphics	22.0	2	28
All other	34.0	2	8
TOTAL		15	106

a   No sampling performed; U.S. coefficients used.

Table 15. Some Wastewater Characteristics of the Textile Industry Subcategory, Five-Commune Area, 1974

Plant Identification Number	Total wastewater discharge, 10^3 m^3	Wastewater discharge per ton of product output, m^3/ton	Percent of total wastewater discharge by subcategory
402	6.07	33.1	1.0
403	31.4	52.3	5.2
404	175.2	136.9	28.8
405	16.9	96.4	2.8
406	9.49	19.4	1.6
407	35.8	16.2	5.9
408	208.0	70.1	34.2
409	27.4	5.4	4.5
410	47.7	139.1	7.8
604	0.873	14.9	.1
411	49.1	80.3	8.1
			100.0

The approach described for industrial activities reflects the fact that, except for twelve outlying industrial plants, all industrial activities discharge into the municipal sewer system. No on-site liquid residuals modification currently takes place, nor is any contemplated. Therefore information on discharges from individual industrial plants with the twelve exceptions noted, is only used to determine quantities discharged to the municipal sewer system.

Estimating Residential BOD_5 and TSS Generation Coefficients. BOD_5 and TSS generation coefficients for residences were estimated by sampling wastewater concentrations in collector lines downstream from isolated, homogenous single-flat and multi-flat residential neighborhoods. The single-flat neighborhood was an old established one; the multi-flat neighborhood was a newly built multi-flat complex. Both locations were sampled at two-hour intervals for 24 hours, on a weekday during a dry period.

Wastewater discharges for the two neighborhoods were estimated from water bills. The number of inhabitants in the single-flat residentail area was obtained from census tract data. For the multi-flat complex, a door-to-door survey was used, because the complex had been constructed subsequent to the most recent census year, 1971. BOD_5 and TSS generation coefficients were then obtained by dividing the total BOD_5 and TSS generated per day by the number of inhabitants in each neighborhood.

Estimating Gaseous Residuals Generation Coefficients

Because essentially no gaseous residuals modification facilities were in place at individual activities or enterprises, except for an electrostatic precipitator at the Moste coal-fired power plant, gaseous residuals generation is the same as gaseous residuals discharge. Generation coefficients for SO_2 and particulates were estimated for all activities except for vehicular travel, which was assumed to be a negligible generator of SO_2 and particulates. Generation coefficients for CO, HC, and NO_x were estimated for all activities except transportation services, for which subcategory generation of these residuals was estimated to be negligible.

Estimating SO_2, Particulates, CO, HC, and NO_x Generation Coefficients for Stationary Sources. Fuel usage was used to identify generators of these five gaseous residuals. Energy conversion was estimated to be the primary source of SO_2 and particulates, and a significant source of HC and NO_x. This assumption results in lower than actual SO_2 and particulates residuals generation coefficients for the pulp and paper subcategory in particular, and probably for SO_2, particulates, and HC for the chemical subcategory, because of excluding the generation of these residuals in production processes. How much lower is unknown because of lack of information on generation of residuals in other than energy conversion. Based on the fuel usage criterion, the activity subcategories listed in tables 7 and 8 were regrouped to form the following "fuel usage" categories: metal processing industry; chemical industry; pulp and paper industry; all other industries; multi-flat residences; single-flat residences; power plants; and all other activities (excluding vehicular movements).

The generation coefficients for SO_2 and particulates used in the study for these fuel usage categories were based on U.S. data.[28] Yugoslav energy conversion technologies were matched with U.S. technologies.[29] Then the corresponding U.S. coefficients were used with the sulfur and ash contents of Yugoslav fuels, which contents were obtained directly from the suppliers. Elementary composition analyses for solid fuels for the years 1970, 1971 and 1972 were provided by the Tribovje and Velenje mines in Slovenia. Average sulfur and ash contents of solid fuels coming from Bosnia were provided by the Banovici mine. The compositions of liquid fuels were obtained from their suppliers: Petrol, Istra Benz, Kurivo, and Kurivoprodaja.

---

[28]The U.S. data are contained in U.S. Department of Health, Education, and Welfare, Rapid Survey Technique for Estimating Community Air Pollution (Cincinnati: Robert A. Taft Sanitary Engineering Center, 1966).

[29]Individuals and organizations consulted for matching energy conversion technologies were: Professor Leopold Andree, Strojna Fakulteta; Michael Fruden, dipl. inz., Strojna Fakulteta; Alojz Hegedic, dipl. inz., Republiski Inspektor Parnih Kotlov; Krusec, dipl. inz., Toplarna Moste; Cizman, Toplarna Siska; Ciuha, dipl. inz., Komunalna Energetika; and Stanovanjsko Podjetje, Fond.

The procedure was as follows. Most industrial boilers--those with steam pressure greater than 0.5 atmospheres and/or water temperature greater than 110°C--are registered by the Slovenian Republic Inspectorate for Boilers, which was able to supply the necessary data for the units registered. Limited data on smaller ordinary household units were provided by the municipal Fire Protection Office. Housing maintenance companies in the area also supplied needed data on the characteristics of energy conversion units used in commercial, institutional, and multi-flat residential activities.

The coefficients for the various energy conversion technologies were then weighted according to the capacities of the energy conversion units in each fuel usage category. The method is illustrated for generation of particulates by the metal processing subcategory, for which the distribution of energy conversion units is shown in table 16.

Table 16. Distribution of Types of Energy Conversion Units in the Metal Processing Fuel Usage Category, Five-Commune Area, 1972

Energy Conversion Technology	Particulate generation coefficient, kg/metric ton burned	Capacity, metric tons of steam/hr	Percent of total capacity
Cyclone	1.0A[a]	1.5	5.4
Hand fired	10.0	0.6	2.2
All other stokers	2.5A[a]	24.5	92.4
TOTALS		26.6	100.0

a A = ash content in percent

The weighted particulate generation coefficient for the metal processing fuel usage category is:

$$(1.0 \text{ A} \times 0.054) + (10.0 \times 0.022) + (2.5 \text{ A} \times 0.924)$$

$$= 0.05 \text{ A} + 0.22 + 2.3 \text{ A};$$

$$= 0.22 + 2.36 \text{ A}.$$

For 1972 the average ash content of the brown coal used was 12.6 percent. Substituting this value for A in the above equation, the resulting particulate generation coefficient for brown coal used in 1972 was 30.0 kg/ton of brown coal combusted. In a like manner, weighted SO_2 and particulates generation coefficients were estimated for each fuel usage category for each type of fuel: heavy oil, light oil, brown coal, lignite, and coke. These are shown in table 17 for the base year, 1972.

Generation coefficients for CO, HC, and NO_x were taken directly from U.S. data[30] in relation to type of activity, size of energy conversion unit, and type of fuels used. Because modeling of air quality in relation to concentrations of CO, HC, and NO_x was not undertaken and no physical measures were directed specifically at reducing discharges of these residuals from stationary sources, CO, HC, and NO_x generation coefficients were not estimated as rigorously as the SO_2 and particulates coefficients.

---

[30]The U.S. data are contained in U.S. Department of Health, Education, and Welfare, Rapid Survey Technique for Estimating Community Air Pollution (Cincinnati: Robert A. Taft Sanitary Engineering Center, 1966).

Table 17. Weighted SO_2 and Particulates Generation Coeficients for Fuel Usage Categories, Five-Commune Area, 1972

(All values in kilograms/metric ton.)

Fuel Usage Category	Fuel Type				
	Heavy liquid fuels[a]	Light liquid fuels[a]	Brown coal	Lignite	Coke
Metal processing	58.9/4.00[b]	19.6/4.00	39.9/30.0	24.7/26.9	19.6/23.8
Chemical	58.9/4.00	19.6/4.00	39.9/30.3	24.7/27.2	19.6/24.1
Pulp and paper	58.9/4.00	19.6/4.00	39.9/24.0	24.7/27.8	19.6/18.7
All other	58.9/4.08	19.6/4.08	39.9/24.7	24.7/14.0	19.6/13.3
Multi-flat	58.9/5.08	19.6/5.06	39.9/10.0	24.7/10.0	N.R.
Single-flat	58.9/6.00	19.6/6.00	39.9/10.0	24.7/10.0	N.R.
Power plants	58.9/4.30	N.R.	37.4/7.50A[c]	24.7/7.50A[c]	N.R.
All other	58.9/5.06	19.6/5.06	39.9/10.0	24.7/10.0	N.R.

Abbreviation: N.R., not relevant, e.g., these fuel types not used by the fuel usage category.

a. Heavy liquid fuel refers to distillate fuel oils. Light liquid fuel refers to residual fuel oil. Distillate fuel oils have a higher sulfur content than residual fuel oils and are much more viscous and more difficult to burn properly. Light liquid fuels have a higher energy content than heavy liquid fuels, ⁻0,000 kcal/kg comparted to 9,600 kcal/kg.

b. Number on left is SO_2 coefficient, number on right is particulates coefficient.

c. A refers to ash content in percent. This coefficient pertains only to the Moste power plant. A weighted coefficient could not be computed for this plant for 1972 because of the many fuel changes which occurred throughout the year, involving various mixtures of lignite and brown coal.

Estimating CO, HC, and NO_x Generation Coefficients for Mobile Sources.
Residuals generation coefficients for vehicular travel are a function of
engine size, fuel type (gasoline or diesel), and velocity, and are expressed as a quantity of residual emitted per vehicle kilometer traveled
(Vkt). Velocity in turn is a function of type of road and traffic conditions. Of the generation coefficients for vehicular travel available
in the literature, those which most closely approximated conditions in
the five-commune area were developed for the engine type/size distribution
in the Federal Republic of Germany.[31] These German coefficients were
adjusted for Yugoslav fuel characteristics. The road network in the
five-commune area was subdivided into the seven road classifications shown
in table 18, for each of which classifications a mean velocity was obtained.[32] Based on these data, CO, HC, and NO_x generation coefficients
were estimated.

Table 18. Classification of Roads in Five-Commune Area

URBAN

Urban interregional (magistrat): sections of arterial highways which
connect regions within Yugoslavia, and Yugoslavia with other
countries, and are in the urban area

Urban regional: sections of main highways which connect points within a region and are in the urban area

---

[31]Anonymous. Emission Kataster Koln (Emissions Inventory of Koln)
(Koln: Ministry for Work, Health, and Welfare, Nordrhein Westfalen,
Federal Republic of Germany, 1970).

[32]American-Yugoslav Project in Regional and Urban Planning Studies,
Spatial Policies for Regional Development, Technical Report No. 8,
(Ljubljana, Yugoslavia: Urbanisticni Institut, 1968).

Table 18 (continued)

URBAN (contd.)

High traffic urban communal: roads within the urban area which carry
heavy traffic volumes, e.g., roads used by commuters

Low traffic urban communal: roads within the urban area which carry
light traffic volumes

NONURBAN

Nonurban interregional (magistrat) roads: sections of arterial highways which connect regions within Yugoslavia, and Yugoslavia with
other countries, and are outside the urban area but within the five-
commune area

Nonurban regional: sections of main highways which connect points
within a region and are outside the urban area but within the
five-commune area

Nonurban communal: all other roads outside the urban area but within the five-commune area not in any of the above classifications.

Source of road classification data: Cestno Padjetje Ljubljana (Ljubljana
Road Company), Cestno Podjetje Kranj (Kranj Road Company), and Republiska
Skupnost za Cesta (Socialist Republic of Slovenia Council of Roads).

Summary of Residuals Generation Coefficients

Table 19 displays the residuals generation coefficients for the

eight residuals of interest for the thirty activity subcategories for the

base year 1972. The accuracy of the estimates of residuals generation

coefficients is a function largely of the homogeneity of the units within

each activity subcategory, and the degree of variation in generation over

time in each unit of the subcategory. For example, with respect to the

former, the consumption patterns of family units living in multi-flat

Table 19. Estimated Mean Residuals Generation Coefficients for Activity Subcategories, Five-Commune Area, 1972

Activity Category and Subcategory	Mixed Solid Residuals		Liquid Residuals					
	Quantity generated per unit		Wastewater Discharge		BOD_5		TSS	
			Quantity generated per unit		Quantity generated per unit		Quantity generated per unit	
	kg	Unit	m^3	Unit	kg	Unit	kg	Unit
INDUSTRIAL								
Metal processing	91	Ton of P.O.	15	Ton of P.O.	0.2	Ton of P.O.	23.9	Ton of P.O.
Electrical products	198	"	25	"	7.1	"	5.5	"
Chemical	63	"	21	"	2.4	"	3.7	"
Building materials	5	"	0.5	"	Included in All other		Included in All other	
Pulp and paper	155	"	150	"	74.2	Ton of P.O.	13.5	Ton of P.O.
Paper products	14	"	1	"	N	"	N	"
Textile	115	"	53	"	7.0	"	5.4	"
Food products	62	"	21	"	19.0	"	8.1	"
Graphics	51	"	16	"	0.2	"	0.5	"
All other	149	"	4	"	3.0	"	2.6	"
RESIDENTIAL								
Multi-flat	0.3	Inhab/day	0.14	Inhab/day	0.07	Inhab/day	0.06	Inhab/day
Single-flat	0.7	"	0.15	"	0.07	"	0.06	"
COMMERCIAL								
Restaurants	858	Emp/yr	0.54	Emp/day	0.07	Emp/day	0.06	Emp/day
Hotels and motels	705	"	0.81	"	0.07	"	0.06	"
Retail stores	344	"	0.20	"	0.07	"	0.06	"
Offices	165	"	0.11	"	0.07	"	0.06	"
Wholesale stores and warehouses	129	"	0.09	"	0.07	"	0.06	"
INSTITUTIONAL								
Government	113	Emp/yr	0.11	Emp/day	0.07	Emp/day	0.06	Emp/day
Research	71	"	0.32	"	0.07	"	0.06	"
Hospitals	159	"	0.05	Patient/day	0.07	Patient/day	0.06	Patient/day
Cultural institutions	261	"	0.32	Emp/day	0.07	Emp/day	0.06	Emp/day
University facilities	45	"	0.16	"	0.07	"	0.06	"
Schools other than university	70	Student/yr	0.02	Student/day	0.07	Student/day	0.06	Student/day
TRANSPORTATION								
Vehicular travel								
Urban	N	10^3Vkt	N	10^3Vkt	N	10^3Vkt	N	10^3Vkt
Nonurban	N	"	N	"	N	"	N	"
Services								
Maintenance	230	Emp/yr	0.55	Emp/day	N	Emp/day	N	Emp/day
Gasoline sales	400	"	0.55	"	N	"	N	"
Offices	165	"	0.11	"	0.07	"	0.06	"
POWER PLANTS								
Moste plant (Solid fuel)	42	10^6cal	41	10^6cal	N	10^6cal	N	10^6cal
Siska plant (Liquid fuel)	N	"	Included with Moste plant		N	"	N	"
COLLECTIVE RESIDUALS HANDLING AND MODIFICATION [a]								
Municipal solid residuals collection agency	165	Emp/yr	0.11	Emp/day	0.07	Emp/day	0.06	Emp/day
Municipal sewerage agency [b]	165	"	0.11	"	0.07	"	0.06	"
Salvaging operations	165	"	0.11	"	0.07	"	0.06	"

Abbreviations: cal, calorie; Emp, employee; Inhab, inhabitant; kg, kilogram; m^3, cubic meter; N, negligible; P.O., product output; Vkt, vehicle kilometers traveled.

a Coefficients relate to office activities.
b $3.0 \times 10^3 m^3$/day of wet solids and 0.02 m^3/day of methane are generated in sewage treatment.

continued...

Table 19. (continued)

Activity Category and Subcategory	Gaseous Residuals									
	SO_2		Particulates		CO		HC		NO_x	
	Qty. gen. per unit		Qty. gen. per unit		Qty. gen. per unit		Qty. gen. per unit		Qty. gen per unit	
	kg	Unit	kg	Unit	kg	Unit	kg	Unit	kg	Unit
INDUSTRIAL										
Metal processing	8.1	Ton of P.O.	7.2	Ton of P.O.						
Electrical products	12.0	"	2.5	"						
Chemical	6.3	"	2.1	"						
Building materials	0.6	"	0.1	"						
Pulp and paper	30.0	"	1.8	"						
Paper products	Included in Pulp and paper		Included in Pulp and paper		a		a		a	
Textile	23.8	Ton of P.O.	0.4	Ton of P.O.						
Food products	5.7	"	0.1	"						
Graphics	3.1	"	0.7	"						
All other	1.0	"	0.2	"						
RESIDENTIAL										
Multi-flat	15.7	Inhab/yr	4.3	Inhab/yr	a		a		a	
Single-flat	20.7	"	4.2	"						
COMMERCIAL										
Restaurants										
Hotels and motels										
Retail stores										
Offices										
Wholesale stores and warehouses	52.9	Emp/yr	13.4	Emp/yr	a		a		a	
INSTITUTIONAL										
Government										
Research										
Hospitals										
Cultural institutions										
University facilities										
Schools other than university										
TRANSPORTATION										
Vehicular travel (Road categories)										
Urban interregional	N	10^3Vkt	N	10^3Vkt	49.3	10^3Vkt	1.8	10^3Vkt	1.0	10^3Vkt
Urban regional	N	"	N	"	19.6	"	0.8	"	1.1	"
High traffic urban communal	N	"	N	"	29.9	"	1.1	"	1.1	"
Low traffic urban communal	N	"	N	"	19.6	"	0.8	"	1.1	"
Nonurban magistrat	N	"	N	"	15.5	"	1.6	"	1.1	"
Nonurban regional	N	"	N	"	17.5	"	0.6	"	1.1	"
Nonurban communal	N	"	N	"	29.9	"	1.1	"	1.1	"
Services										
Maintenance										
Gasoline sales	Included in COMMERCIAL		Included in COMMERCIAL		N	"	N	"	N	"
Offices										
POWER PLANTS										
Moste plant (Solid fuel)	1.0	10^6cal	0.9	10^6cal	a		a		a	
Siska plant (Liquid fuel)	3.5	"	N	"						
COLLECTIVE RESIDUALS HANDLING AND MODIFICA-TION										
Municipal solid residuals collection agency										
Municipal sewerage agency	Included in COMMERCIAL		Included in COMMERCIAL		Included in COMMERCIAL		Included in COMMERCIAL		Included in COMMERCIAL	
Salvaging operations										

a Coefficients based on type of fuel used as follows:
   Liquid fuel: CO, 0.0049 grams liter (g/l); HC, 0.38 g/l; NO_x, 12.4 g/l

   Solid fuel: CO, 25.0 kilograms/ton (kg/t); HC, 5.0 kg/t; NO_x, 4.0 kg/t

residences in the five-commune area appear to be quite similar. Therefore, the limited sampling of MSR generation performed at multi-family residences is believed to have produced reasonable MSR generation coefficients. On the other hand, production activities can vary significantly among plants in the same industrial activity subcategory. Consequently, residuals generation coefficients based on sampling at only two or three plants in a broadly defined industrial activity subcategory are likely to be less accurate. The typically large variation in residuals generation from day-to-day in an industrial plant, and in some institutional and commercial activities, makes it difficult to obtain a representative sample from which to derive a residuals generation coefficient.

Three other factors merit mention. One, it is not always easy to locate all discharge points in many industrial plants, particularly for old plants. Two, production records are not always maintained in a manner such that the quantities of product outputs during the sampling periods can easily be identified. Rarely does a plant operate continuously at the same output level with the same product mix unless it is a relatively simple one raw material/one process/one product plant. Three, data on residuals generation and on product output are not always available for the same year.

Levels of Activity and Residuals Generation

Having estimated relevant residuals generation coefficients for each of the activity subcategories, the level of activity for each subcategory was obtained directly from individual activities (such as the Moste power plant), from census data, or from special studies. For example, the

number of Vkts traveled in each of the seven road categories by diesel
and gasoline fueled vehicles was estimated by multiplying the 1972 traffic
count[33] for each type by the length of road in that type. Table 20 shows
the generation of residuals in the five-commune area by activity subcategory, for 1972. Table 20 shows clearly which subcategories are the
principal generators of each residual.

The next step in the REQM analysis was to estimate for the five-
commune area the costs of various physical measures for residuals modifica-
tion--collection, transport, modification, and disposal--in relation to the
environmental quality targets, e.g., 80 percent reduction in discharge of
BOD_5. The procedures for estimating these costs are discussed in the next
chapter.

---

[33]The traffic count data, obtained from Republisko Skupnost za Costa
SRS (Republic Committee for Roads), were adjusted for seasonal variations.

Table 20. Estimated Residuals Generation by Activity Subcategory, Five-Commune Area, 1972

Activity Category and Subcategory	Mixed Solids Residuals Thousand metric tons	% of area total	Liquid Residuals Wastewater Discharge Million cubic meters	% of area total	BOD Hundred metric tons	% of area total	TSS Hundred metric tons	% of area total
INDUSTRIAL								
Metal processing	8.6	5.0	1.5	1.7	0.2	0.1	4.5	3.7
Electrical products	3.9	2.2	0.5	0.6	1.4	0.8	1.1	0.9
Chemical	6.2	3.6	2.1	2.4	2.4	1.3	3.7	3.0
Building materials	4.8	2.8	0.5	0.6	Included in All other		Included in All other	
Pulp and paper	9.2	5.3	9.1	10.3	44.0	24.2	8.0	6.5
Paper products	0.4	0.2	N	N	N	N	N	N
Textile	1.3	0.7	0.6	0.7	0.8	0.4	0.6	0.5
Food products	10.6	6.1	3.7	4.2	32.3	17.8	14.0	11.4
Graphics	1.3	0.7	0.4	0.5	0.1	N	0.1	N
All other	8.7	5.0	0.2	0.2	1.5	0.8	1.5	1.2
TOTAL INDUSTRIAL	55.0	31.6	18.6	21.2	82.7	45.5	33.5	27.3
RESIDENTIAL								
Multi-flat	19.4	11.2	8.1	9.1	42.1	23.2	38.0	30.9
Single-flat	21.3	12.3	3.3	3.7	21.4	11.8	19.3	15.7
TOTAL RESIDENTIAL	40.7	23.5	11.4	12.8	63.5	35.0	57.3	46.6
COMMERCIAL								
Restaurants	2.2	1.3	0.5	0.6	0.5	0.3	0.4	0.3
Hotels and motels	0.7	0.4	0.3	0.3	0.2	0.1	0.2	0.2
Retail stores	2.3	1.3	0.5	0.6	1.4	0.8	1.3	1.0
Offices	1.5	0.9	0.4	0.5	1.9	1.0	1.7	1.4
Wholesale stores and warehouses	1.6	0.9	0.4	0.5	2.6	1.4	2.3	1.9
TOTAL COMMERCIAL	8.3	4.8	2.1	2.5	6.6	3.6	5.9	4.8
INSTITUTIONAL								
Government	0.4	0.2	0.1	0.1	0.8	0.4	0.7	0.6
Research	0.2	0.1	0.4	0.5	0.7	0.4	0.6	0.5
Hospitals	0.8	0.5	1.4	1.6	15.9	8.8	14.5	11.8
Cultural institutions	0.3	0.2	0.1	0.1	0.2	0.1	0.1	N
University facilities	0.1	N	0.1	0.1	0.2	0.1	0.4	0.3
Schools other than university	3.4	2.0	0.3	0.3	9.9	5.5	9.0	7.3
TOTAL INSTITUTIONAL	5.2	3.0	2.4	2.7	27.7	15.3	25.3	20.6
TRANSPORTATION								
Vehicular travel								
Urban	N	N	N	N	N	N	N	N
Nonurban	N	N	N	N	N	N	N	N
Services								
Maintenance	3.2	2.0	0.8	0.9	N	N	N	N
Gasoline sales	0.1	N	N	N	N	N	N	N
Offices	0.3	0.2	0.1	0.1	0.5	0.3	0.4	0.3
TOTAL TRANSPORTATION	3.6	2.2	0.9	1.0	0.5	0.3	0.4	0.3
POWER PLANTS								
Moste plant (Solid fuel)	60.5	34.9	52.7	59.4	0.1	N	0.1	N
Siska plant (Liquid fuel)	N	N	Included in Moste plant					
TOTAL POWER PLANTS	60.5	34.9	52.7	59.4	0.1	N	0.1	N
COLLECTIVE RESIDUALS HANDLING AND MODIFICATION (CRHM)								
Municipal solid residuals collection agency	0.1	N	N	N	N	N	N	N
Municipal sewerage agency[a]	N	N	0.6	0.7	0.5	0.3	0.4	0.3
Salvaging operations	0.1	N	N	N	N	N	N	N
TOTAL CRHM	0.2	N	0.6	0.7	0.5	0.3	0.4	0.3
FIVE-COMMUNE AREA TOTALS	173.5	100	88.7	100	190.5	100	124.3	100

Abbreviation: N, negligible.

a 1.7×10^6 cubic meters of wet solids (sludge) and 7.3 cubic meters of methane are estimated quantities generated each year.

continued...

Table 20. (continued)

Activity Category and Subcategory	SO₂		Particulates		CO		HC		NO$_x$	
	Thousand metric tons	% of area total	Thousand metric tons	% of area total	Thousand metric tons	% of area total	Thousand metric tons	% of area total	Thousand metric tons	% of area total
INDUSTRIAL										
Metal processing	0.8	3.0	0.7	12.7						
Electrical products	Included in All other		Included in All other							
Chemical	0.6	2.2	0.2	3.6						
Building materials	Included in All other		Included in All other							
Pulp and paper	2.2	8.1	1.3	23.6	Estimated for INDUSTRIAL Activity Category as a whole					
Paper products	Included in Pulp and paper		Included in Pulp and paper							
Textile	Included in All other		Included in All other							
Food products	"		"							
Graphics	"		"							
All other	2.3	8.5	0.4	7.3						
TOTAL INDUSTRIAL	5.9	21.8	2.6	47.6	0.2	0.1	0.1	5.8	1.4	17.1
RESIDENTIAL										
Multi-flat	2.0	7.4	0.5	9.1	Estimated for RESIDENTIAL Activity Category as a whole					
Single-flat	1.7	6.3	0.4	7.3						
TOTAL RESIDENTIAL	3.7	13.7	0.9	16.4	1.9	7.3	0.4	23.5	0.7	8.5
COMMERCIAL										
Restaurants	Estimated for COMMERCIAL, INSTITUTIONAL, TRANSPORTATION, and COLLECTIVE RESIDUALS HANDLING AND MODIFICATION Activity Categories combined									
Hotels and motels										
Retail stores										
Offices										
Wholesale stores and warehouses										
TOTAL COMMERCIAL	2.7	10.0	0.7	12.8	1.6	6.2	0.3	17.6	0.3	3.7
INSTITUTIONAL										
Government										
Research										
Hospitals										
Cultural institutions										
University facilities										
Schools other than university										
TOTAL INSTITUTIONAL										
TRANSPORTATION										
Vehicular travel										
Urban	N	N	N	N	18.6	71.3	0.7	41.5	0.5	6.1
Nonurban	N	N	N	N	3.6	13.8	0.1	5.2	0.2	2.4
Services										
Maintenance										
Gasoline sales	Included with COMMERCIAL, INSTITUTIONAL, and COLLECTIVE RESIDUALS HANDLING AND MODIFICATION Activity Categories									
Offices										
TOTAL TRANSPORTATION					22.2	85.1	0.8	47.3	0.7	8.5
POWER PLANTS										
Moste plant (Solid fuel)	14.6	53.8	1.3	23.6	Estimated for POWER PLANTS Activity Category as a whole					
Siska plant (Liquid fuel)	0.2	0.7	N							
TOTAL POWER PLANTS	14.8	54.5	1.3	23.6	0.1	N	0.1	5.8	5.1	62.2
COLLECTIVE RESIDUALS HANDLING AND MODIFICATION (CRHM)										
Municipal solid residuals collection agency	Included with COMMERCIAL, and INSTITUTIONAL Activity Categories, and TRANSPORTATION Services Activity Subcategory									
Municipal sewerage agency										
Salvaging operations										
TOTAL CRHM										
FIVE-COMMUNE AREA TOTALS	27.1	100	5.5	100	26.1	100	1.7	100	8.2	100

Chapter VI

ACTIVITY MODELING: ESTIMATING RESIDUALS
HANDLING AND MODIFICATION COSTS

Introduction

In any given REQM context, the least cost combination of physical
measures to achieve E.Q. targets is not likely to be found unless the
entire range of physical measures enumerated in chapter II for improving
AEQ is analyzed. Factors which must be considered in selecting physical
measures for analysis in any given case include: the nature of the production processes and raw materials used; the financial and labor resources
available for installation and operation; the existence of alternative
technological options from foreign sources; possible barriers to the purchase of foreign technological options; the extent to which a particular
physical measure is considered politically, socially, economically feasible, e.g., can be implemented; and the available analytical resources.
For the Ljubljana study, the last two factors in particular limited the
physical measures analyzed. For example, it appeared doubtful that such
measures as changing manufacturing production processes or family consumption patterns would be considered feasible, even if data on such possibilities could have been generated.

The costs of residuals modification, handling, and disposal are costs
which are incurred by an activity in producing a product or service or in
carrying on the activities within the activity unit, e.g., within a
dwelling unit. These costs are in addition to normal production or

activity costs, and would not be incurred except for constraints imposed
on residuals discharges in order to achieve E.Q. targets. Thus, the proper
calculation of residuals modification costs ("pollution control" costs)
involves the comparison of the costs which would be incurred over time
without pollution controls with the costs over the same time period with
the controls.

Four additional points with respect to costs merit emphasis. First,
in some cases the measures installed result in recovery of some usable
materials and energy, or otherwise reduce operating costs. Whatever
savings are achieved must be subtracted from the gross costs, in order
to obtain the net costs of residuals modification. If the measures increase other operating costs, the additional costs must be included, and
attributed to residuals modification.

Second, many physical measures can be installed at different levels
of intensity. For example, a cyclonic filter for installation on a boiler
stack could be designed to remove 80 percent, 90 percent, 95 percent,
of the particulates from the discharge stream.

Third, many physical measures which modify residuals after generation
result in the generation of secondary residuals, which in turn must be
handled and modified and/or disposed of to the environment. For example,
the particulates removed by a cyclonic filter become fly ash which may be
placed in a landfill.

Fourth, conceptually in REQM analyses, measures to reduce residuals
discharges would be selected and analyzed in relation to each discharger,

so that possible physical measures would be developed in relation to the specific conditions of that discharger. Even in the same industrial or other activity subcategory, the least cost combination of measures to achieve a given level of discharge reduction would likely be different for each of the units in the activity subcategory. In the Ljubljana study, the available analytical resources did not permit differentiation of types of physical measures among the different units in an activity subcategory, except for liquid residuals modification in twelve outlying industrial plants. That is, the same physical measure at the same level of intensity was applied to each unit in an activity subcategory. However, the costs of the application do reflect the size of each unit and the particular technology existing in each unit.

The costs for each measure applied to an activity include: initial capital costs, e.g., site preparation, new equipment, installation, and annual operating and maintenance (O&M) costs, e.g., fuel, electrical energy, labor, maintenance materials. For equipment which has a physical life less than the time period of analysis, replacement costs must be included, as discussed in the final section of this chapter dealing with present value calculations. All costs are in 1974 New Dinars (N.D.). Associated with each physical measure at its level of intensity of application is a cost, the reduction in residuals discharge achieved, and—where relevant—the impact on AEQ. The last is discussed in chapters VII and VIII.

The development of residuals modification costs is discussed herein
for: (1) reduction in SO_2 and particulate discharges from stationary
sources; (2) reduction in CO, HC, and NO_x from mobile sources; (3) reduction in BOD_5 and TSS from point sources; and (4) handling and disposal
of MSR.

Estimating Costs of Reducing SO_2 and Particulate Discharges from Stationary Sources

Industrial Sources

Three measures to reduce SO_2 discharges from enterprises in the
industry category were analyzed: (1) conversion of boilers using coal to
boilers using light liquid fuel; (2) conversion of boilers using heavy
liquid fuel to boilers using light liquid fuel; and (3) desulfurization
of heavy liquid fuel for use in boilers using heavy liquid fuel. For
analysis of the first two measures, the following information was obtained
on all boilers used in industry: boiler technology; type of fuel used;
capacity; and age of boiler. Based on this information Yugoslav engineers
estimated the capital costs--equipment purchase and installation--for
modifying or installing burners and storage tanks. Energy conversion
efficencies for each boiler were estimated for present installations and
the alternative installations. The cost of SO_2 reduction then was estimated as the difference between the present value of the time stream of
costs of using the light liquid fuel and the time stream of costs of
obtaining the same quantity of energy from the currently used coal or

heavy liquid fuel. Additional maintenance costs were estimated to be negligible, which is very likely a conservative estimate. Both light liquid fuels and desulfurized heavy liquid fuels may decrease boiler scaling and corrosion. The elimination of coal eliminates the cost of ash removal and disposal.

With respect to the third option, desulfurization, it was assumed that a desulfurization plant would be constructed outside of the five-commune area to provide desulfurized heavy liquid fuel for use not only in the Ljubljana area but in other areas of the country as well. Based on the estimated total demand for the output of such a plant, Yugoslav engineers estimated the average cost of the desulfurized fuel produced by a plant of that size. This unit cost was then used as the unit price for the amount of desulfurized heavy liquid fuel necessary to generate an equivalent amount of energy in the relevant boilers in the five-commune area.

Reductions in SO_2 discharges were the estimated difference between SO_2 discharges under present conditions and SO_2 discharges with one or more of the alternative measures in place. The capital and O&M costs and the corresponding reductions in SO_2 discharges for various physical measures applied to the metal processing industry fuel usage category are shown in table 21. Costs for various physical measures to reduce SO_2 discharges were developed in a similar manner for three other industrial fuel usage categories--pulp and paper, chemical, and all other.

Table 21. Estimated Costs[a] of and Percent Reduction in SO_2 Discharges from the Metal Processing Industry in the Five-Commune Area, 1972 Conditions

Alternative	Capital Costs 10^6 N.D.	O&M Costs 10^6 N.D./Yr.	Percent Reduction in Discharge of SO_2 from Metal Processing
I Convert boilers using coal[b] to light liquid fuel units	2.7	3.2	74.0
II Desulfurize heavy liquid fuel for boilers using heavy liquid fuels[c]	None[d]	0.5	5.6
III Convert boilers using heavy liquid fuels to light liquid units[c]	0.2	0.5	4.5
I plus II	2.7	3.7	79.6[e]
I plus III	2.9	3.7	78.5[e]

a Costs are in 1974 N.D.
b Includes all solid fuels.
c Alternatives II and III are mutually exclusive.
d The capital costs of a desulfurization plant are not shown as a capital cost for the metal processing industry. They are reflected in higher fuel prices paid by the industry and therefore are included in O&M.
e For the region, percent reductions in regional SO_2 discharges in applying alternatives I plus II and I plus III to the metal processing industry would be about 2.2 percent and 2.1 percent, respectively.

Each physical measure applied to reduce the discharge of SO_2 also affects, to varying degrees, the discharges of other gaseous residuals from these activities. For example, converting existing coal-using boilers to light liquid fuel boilers in the metal processing industry to reduce SO_2 discharges results in a 74 percent reduction in SO_2 discharges from the industry and simultaneous reductions in discharges of particulates, CO, HC, and NO_x of about 80 percent, 99 percent, 84 percent, and 74 percent, respectively.

The alternatives for reducing discharges of particulates from industrial enterprises are: (1) converting boilers using coal to boilers using light liquid fuel; and (2) installation of cyclonic filters on stacks of plants with coal fueled energy conversion units. Costs for the cyclonic filter alternative were estimated based on a 95 percent removal efficiency.

Wet scrubbing and limestone injection were not considered as feasible alternatives to reduce SO_2 and particulate discharges from energy conversion units, because of excessive costs and technical difficulties in obtaining the requisite equipment in Yugoslavia.

Nonindustrial Sources

There are two major alternatives for reducing discharges of SO_2 and particulates from nonindustrial activities. One is conversion of existing energy conversion units fueled with coal or heavy liquid fuel in individual structures or units, e.g., single-flat or multi-flat residences, to units using light liquid fuels. The other is provision of steam for space heating and hot water from a central heating plant or plants rather than from facilities in individual structures or units. Analysis of the first is analogous to the analysis of changing fuel in the industrial activity category. Analysis of the second requires detailed discussion.

During recent years in the five-commune area it had been hypothesized that air quality in the urbanized portion of the area, as measured by concentrations of SO_2 and TSP, could be greatly improved by connecting all new residential, institutional, and commercial structures and some existing structures, in addition to those already connected, to a central heating plant or plants which would supply steam for space heating and hot water. That hypothesis concerning off-site central heating had never been investigated until this study. It was assumed that the types of fuels used in the central plants would be the same as at present, namely, a brown coal/lignite mixture and heavy liquid fuel, because the unpredictable supplies and fluctuating prices of low sulfur solid and liquid fuels in the quantities required would preclude consistent use of such fuels.

Because steam can be transported efficiently only a short distance, not all parts of the urbanized area can be connected to central heating plants. (The additional boiler capacity required for off-site central heating would be constructed on sites of existing power and heating plants.) The fifty-three sectors into which the five-commune area had

been divided, shown in figure 11, were used to determine which parts of
the area could be physically and economically connected. Analysis indicated that connecting fourteen of the fifty-three sectors was feasible.
These fourteen, shown in figure 12, were combined into the following
five connection areas:

Connection Area	Sectors
V	1, 2, 3(1/2), 4
W	17, 18, 19
X	9, 10, 11
Y	3(1/2), 22, 26
Z	20, 21

The fourteen sectors would be served, sequentially, by: (1) connecting
to existing unused boiler capacity at each of the two power plants in the
area; (2) adding additional boiler capacity at the coal-fired plant, to
be used only for steam for space heating and hot water; and (3) adding
additional boiler capacity at the heavy liquid fuel plant to be used
only for steam for space heating and hot water.

The capital and O&M costs,[33] and the associated impacts on discharges of SO_2 and particulates in the region are shown in figures 13A
and 13B, respectively. The sequence of plotting reflects the fact that
the objective is to reduce SO_2 discharges, because the mean ambient concentration of total suspended particulates already meets the standard.

---

[33]The costs were provided by Engineer Solar, Komunalno Podjetje
Energetika, Toplarna, Ljubljana, Yugoslavia, February 1974.

Figure 11. Sectors in the Five-Commune Area

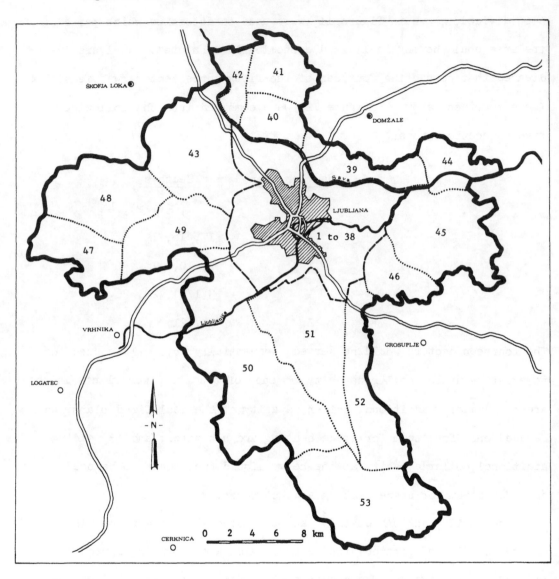

Key: ▬▬▬ Five-commune area boundary   ▬ ▬ ▬ 38 Sectors boundary

══ Main road   •••••• Sector boundary

▬ River   ▨▨▨ Urbanized area

131

Figure 12. Delineation of Feasible Connection Sectors for
Space Heating from Central Plants

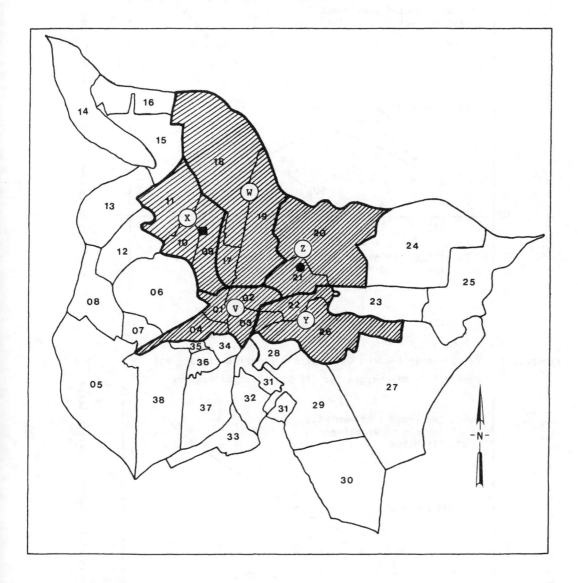

Key: ▨ Feasible connection sectors
      ● Present coal-fueled power plant
      ■ Present central heating plant using heavy liquid fuel

Figure 13A. Relationship between Capital Cost and Reduction in SO_2 and Particulate Discharges for Off-Site Central Heating

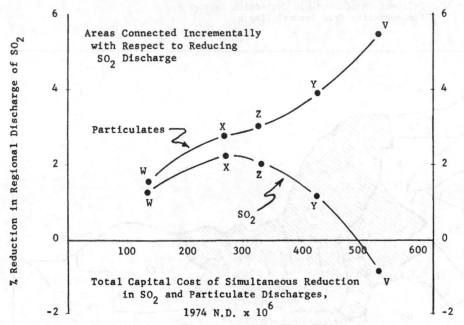

Figure 13B. Relationship between O&M Cost and Reduction in SO_2 and Particulate Discharges for Off-Site Central Heating

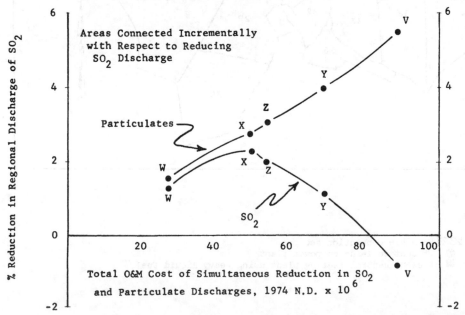

The costs shown are joint costs for the simultaneous changes in discharges of SO_2 and particulates. The impacts on CO, HC, and NO_x discharges are not shown.

If all relevant structures in the five feasible connection areas were connected to central heating plants as proposed, SO_2 discharges in the region would actually increase by about 1 percent rather than decrease as hypothesized. Discharges of particulates in the region would be decreased by only about 5 percent. Three factors account for these results. First, many dwellings using on-site energy conversion units usually heat only one or two rooms. In contrast, where a dwelling unit is served by off-site central heating usually all rooms are heated, thereby requiring more energy and more fuel combustion and resulting in more residuals generation per dwelling unit. Second, most existing on-site energy conversion units use higher energy, less residuals containing fuels--light liquid fuels and brown coal--than those used in the central heating plants, namely, a brown coal-lignite mixture and heavy liquid fuel. To provide an equivalent amount of usable heat energy after conversion to off-site heating would require the use of a greater amount of lower energy, high sulfur fuels by the heating plants. Third, additional heating plant energy must be generated to compensate for significant heat losses in the distribution network for off-site central heating. These three factors together require that off-site central heating convert about 1.6 times the energy converted by conventional on-site heating methods, with consequent increases in total regional SO_2 and particulates generation.

Because there are no measures installed in the heating plants to
reduce SO_2 discharges, SO_2 generation equals SO_2 discharge. A small net
reduction in SO_2 discharge in the five-commune area is achieved when
only areas W and X are connected, as shown in figures 13A and 13B. This
results from the fact that areas W and X are connected to the liquid fuel
plant, which generates smaller quantities of SO_2 and particulates per
unit of energy produced than does the brown coal-lignite fueled plant,
to which all other areas are connected. Extending off-site central
heating to connection areas V, Y, and Z results in a net increase in
regional SO_2 discharges. With respect to discharges of particulates,
electrostatic precipitators at the brown coal-lignite fired plant result
in a small net decrease in discharges of particulates in the five-commune
area, in spite of the increase in generation of particulates at the plant.
In addition, because of the increased total amount of solid fuel combusted under this alternative, the generation and discharge of CO, HC,
and NO_x would also increase.

Estimating Costs of Reducing CO, HC, and NO_x Discharges from Mobile Sources

Vehicular travel is the major source of CO discharges in the region
and a major source of HC discharges. Two alternatives for reducing CO,
HC, and NO_x discharges from mobile sources were investigated. The first,
termed the "park/ride" alternative, had two possible levels of application. The second was the carburetor idling adjustment alternative. The
practicality of this second alternative was demonstrated in Austria in

the ABGASENTIGIFTUG study in 1965. The study analyzed 4,381 vehicles
manufactured by six different producers, and showed that carburetor
idling adjustment could reduce CO emissions by a factor of ten.

The park/ride alternative involves reducing in-bound commuter traffic
to, and out-bound commuter traffic from, the urbanized area, which traffic accounts for about 50 percent of the Vkt in the urbanized area. This
would be achieved by developing bus transportation from outlying commuter
parking lots to the center of the city. Two levels of application, I and
II, were formulated. Level I involves two, and Level II involves four,
parking lots, the locations of which are shown in figure 14. The total
available parking space in the four lots for Level II is about 80 percent
of the total space available in the two lots for Level I. To transport
commuters to the city center Level II requires 1.7 times the number of
buses that Level I requires. However, Level II would result in an
estimated Vkt reduction of about 13 percent compared with an estimated 7
percent reduction with Level I. The reason is that, although fewer cars
can be parked in the four smaller lots than in the two larger lots, the
number of Vkt saved by parking fewer cars farther from the city center
is greater than the number saved by parking more cars nearer to the city.

Table 22 shows the capital and O&M costs for Levels I and II of the
park/ride alternative, and the corresponding estimated reductions in
Vkt and the changes in CO, HC, and NO_x discharges. Capital costs include purchase of land, purchase of buses, and construction of the
parking lots. O&M costs include labor, fuel, maintenance on the buses,

Figure 14. Parking Lot Locations for Levels I and II
of Park/Ride Alternative

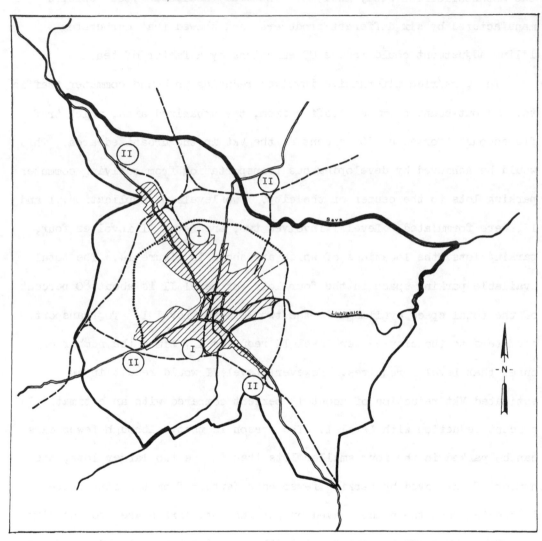

Key:
▬▬▬ Boundary of sectors 1-38	I	Proposed parking lot Alternative I
▭▭▭ Existing main road		
– – – Planned highway	II	Proposed parking lot Alternative II
••••• Planned Ring Road		
	▨▨▨	Urbanized area

Table 22. Estimated Costs[a] of and Percent Reductions in Discharges of CO, HC, and NO_x from Mobile Sources in the Five-Commune Area, 1972 Conditions

Alternative	Capital Costs 10^6 N.D.	O&M Costs 10^6 N.D./ Yr.	Percent Reduction in Vkt	Percent Reduction in Discharge of		
				CO	HC	NO_x
Park/Ride Level I (2 Lots)	260	64	7	11	11	7
Park/Ride Level II (4 Lots)	240	52	13	19	19	11
Carburetor Idling Adjustment[b]	0.9	12.0	--	20	2	--[c]
Park/Ride Level I plus Carburetor Idling Adjustment	261	76	7	29	13	<7
Park/Ride Level II plus Carburetor Idling Adjustment	241	64	13	35	21	<11

a Costs are in 1974 N.D.
b Percent reductions for the carburetor idling adjustment alternative were estimated by Dr. J. Pavletic, Machine Faculty, University of Ljubljana, March 1974.
c NO_x discharge increases with the application of this alternative, but the amount of increase is not known.

and labor at and maintenance of the parking lots. Park/ride alternative Level I is more expensive than Level II because: (1) land costs are higher nearer the city; (2) the total area of the two lots is greater than the total area of the four lots; and (3) the capital and O&M costs of the greater number of additional buses required for Level II are small relative to the costs affected by (1) and (2). Net reductions in CO, HC, and NO_x discharges are achieved by the park/ride alternative because automobile Vkt are reduced more than bus Vkt are increased.

The carburetor idling adjustment alternative involves adjusting, every four to five months, air-fuel mixtures and idling speeds on all gasoline powered vehicles in the five-commune area. Based on data from outside Yugoslavia, such a procedure is estimated to reduce CO and HC discharges significantly. Capital costs of this alternative include the purchase of monitoring and patrol vehicles and exhaust measurement instruments for patrol vehicles and inspection stations. It was estimated that sufficient capacity (space and manpower) exists in the present inspection stations to handle the incremental demand. Maintenance costs include labor for monitoring and inspection, and for adjustment of vehicles. The capital and O&M costs and the corresponding reductions in CO and HC discharges are shown in table 22. A disadvantage of the carburetor idling adjustment alternative is that, although CO and HC discharges are reduced, NO_x discharges are increased. The magnitude of the increase is not known.

Estimating Costs of Reducing BOD_5 and TSS Discharges from Point Sources

BOD_5 and TSS residuals are generated in all activity subcategories in the five-commune area except vehicular travel. These residuals can be modified on-site or can be collected and modified in collective facilities operated by a municipal agency or some group of enterprises. In the five-commune area, the municipal sewerage agency operates the sewer system which collects wastewater from a substantial portion of the five-commune area. The collected wastewater is not modified prior to discharge into the Sava and Ljubljanica rivers. The liquid residuals generated by activities in the remainder of the five-commune area are either discharged directly to the Sava or Ljubljanica rivers or into septic tanks and/or cesspools. Because of the high water table in much of the area, many of the septic tanks and cesspools result in relatively little modification of the liquid residuals. About 4 percent of the population in the five-commune area is served by four small wastewater treatment plants.

Given the existing spatial distribution of activities in the five-commune area and given that no physical measures to reduce generation of BOD_5 and TSS in the various activities were considered possible, the only options for reducing BOD_5 and TSS discharges in the five-commune area are on-site modification and collective modification in wastewater treatment plants. Because of economies of scale, unit costs of treatment are substantially lower in collective facilities than in facilities at individual sites. However, to costs of collective treatment plants must be added collection costs associated with transporting wastewaters from

individual sites where wastewaters are generated to the collective
facility. Depending on collection costs, connecting all or only a portion
of the individual activities in a region to the collective facility will
yield the minimum regional cost for reduction in BOD_5 and TSS discharges.

The first step in developing BOD_5 and TSS discharge reduction costs
was to inventory all activities in each subarea of the five-commune area to
determine which activities were discharging to the sewer system, which
directly to surface waters, and which to septic tanks and/or cesspools.
Based on this inventory and the decision already made by the municipal
sewerage agency with respect to the location for the municipal sewage
treatment plant, it was estimated that it would be economically feasible
to connect almost all activities in the five-commune area to the collective
facility, except for some scattered activities in outlying sections of the
area. The assimilative capacity in the outlying sections is sufficient
so that all activities in these outlying sections, other than twelve industrial enterprises, can discharge into septic tanks, cesspools, or
directly to surface or ground water bodies without adverse impacts on
ambient water quality. Therefore, detailed cost estimates were prepared
for BOD_5 and TSS discharge reduction in a collective wastewater treatment
plant and in on-site plants at each of the twelve outlying industrial
activities.

Estimating Costs of Reducing BOD$_5$ and TSS Discharges
in the Collective Wastewater Treatment Plant

The total cost of any given degree of BOD$_5$ and TSS discharge reduction by a collective facility is a function of: the length and capacity
of the collection network; the characteristics of the area in which it
is to be installed; the capacity and technology of the treatment plant;
and the quantity of liquid residuals to be handled. To find the minimum
cost for BOD$_5$ and TSS discharge reduction involves both network design
and treatment plant design. For the five-commune area, both the location
of the treatment plant and the technology of residuals modification to
be used were selected by the municipal sewerage agency (Kanalizacija).
These decisions provided the starting conditions for the analysis.

Designing the Collection Network. The elements of the collection
network include: (1) the primary (main) collector which usually runs
through the center of a large connection area; (2) the secondary pipes
(laterals) which run from the primary collector to the boundaries of the
activities; and (3) the connector from each activity to a lateral. The
construction cost of the network is a function of geologic and topographic
characteristics, labor costs, material costs, and methods of construction.
Additional capital costs for a separate storm water runoff system were
included where applicable. O&M costs of the collection network were
included as a component of O&M costs for the total liquid residuals collection and modification system.

The collection network analysis was based on the fifty-three sectors
into which the five-commune area had been divided. In designing the network, the spatial distribution of population was superimposed over the
fifty-three sectors (shown in figure 11). Most of the residuals generating
activities are located in sectors one through thirty-eight. The location
of each individual industrial source was plotted. Because of a lack of
specific data on locations, commercial, institutional, and transportation
service activities were assumed to be distributed over the urbanized area
in proportion to population. Combining the locations and levels of activities data with residuals generation coefficients for WW, BOD_5, and
TSS for the respective activities yielded the total loading of wastewater,
BOD_5, and TSS for each sector.

Data on capacities of the existing sewer network indicated that the
existing collection system in each of the sectors presently connected to
the sewer system had sufficient capacity to handle the discharges from all
presently unconnected activities in those sectors. However, not all
sectors need to be connected, because in some the assimilative capacity
is sufficient to permit using on-site septic tank or cesspool disposal,
or no modification at all. Eliminating these sectors yields the net
loading to the sewer collection network. The network design problem then
became that of determining the minimum cost sequence by which sectors
would be connected to the sewage collection network. The criterion for
determining sequence was total load connected per million N.D. of collection system expenditures.

Topographical and geological maps were superimposed over the sectors to determine the direction of drainage and foundation (excavation, construction) conditions. This enabled differentiating construction costs on the basis of different soil and rock characteristics, and slope, in designing the main collector system. Based on all of the above information, nine separate areas were formed from the fifty-three sectors and aggregated into the five connection areas shown in figure 15.

Total collection network cost for each sector was then estimated as follows. Historical data on unit site connection costs--cost per site connection per unit WW flow--for various topographic-geologic conditions were adjusted to 1974 N.D. by Kanalizacija. Similarly, historical data on unit costs for laterals and main collector--cost per hectare--based on soil type, degree of development in the sector, and type of sewer system, e.g., separate or combined, were adjusted to 1974 N.D. The number of new site connections multiplied by the respective unit site connection costs plus the number of hectares containing new laterals and a new main collector multiplied by their respective unit costs comprised the capital cost of the collection network. Annual O&M costs for the network were estimated at 2 percent of the initial capital costs. The present value of network costs per unit of wastewater was then calculated for each of the five connection areas. This ratio comprised the basis for determining the most efficient sequence of connecting the areas to the system, e.g., beginning with the connection area having the lowest additional collection network cost per unit of WW collected.

Figure 15.   Delineation of Sewage Connection Areas
in the Five-Commune Area

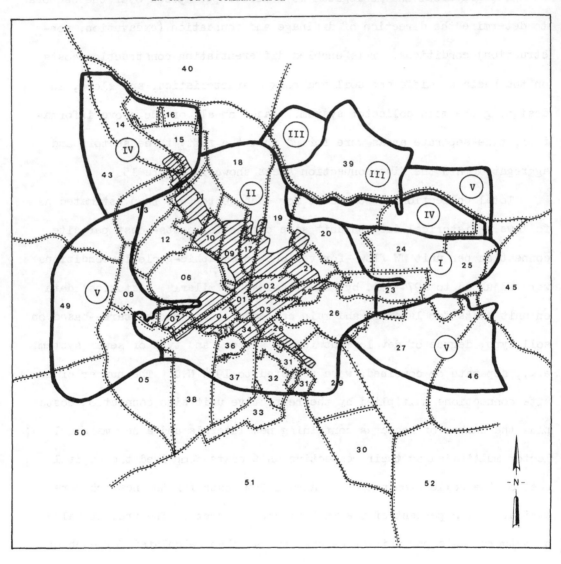

Key: ▬▬▬ Boundary of network extension area
      ┼┼┼┼ Sector boundary
      ▨▨▨ Urbanized area

Designing the Wastewater Treatment Facility. It was difficult to estimate the costs of sewage treatment facilities in Yugoslavia for two reasons. First, only a few sewage treatment facilities of very small capacities, e.g., less than 15,000 population equivalents, had been constructed in Slovenia at the time of the study. Second, there are no manufacturers of sewage treatment facilities in Yugoslavia. Hence, an Austrian firm--Ruthner Industrieanlagen of Vienna--was asked to design and estimate the costs of constructing and operating two activated sludge treatment plants of two different capacities, for the specified site in the five-commune area. Two designs were produced for each capacity, one for 70 percent BOD_5 removal and one for 95 percent BOD_5 removal. The designs explicitly considered: (a) TSS as well as BOD_5; (b) the fact that BOD_5 is generated as well as removed in the activated sludge process; and (c) handling the sludge generated in the process. The design firm was provided with the necessary influent characteristics and site data by Kanalizacija. The final cost estimates were adjusted to consider import fees, exchange rates, the substitution of available Yugoslav components (as required by law), and inflation, and were expressed in 1974 N.D.

To obtain capital and O&M costs for activated sludge plants of other capacities, it was assumed that the linear relationship between the log of plant capacity and the log of unit cost found in the United States[34]

---

[34] Robert Smith, Cost of Conventional and Advanced Treatment of Wastewaters (Cincinnati: Robert A. Taft Sanitary Engineering Center, July 1968).

was valid in Slovenia. This assumption and the cost estimates by the
design firm for two plant sizes for a given level of BOD_5 removal enabled interpolating for and extrapolating to other capacities for the
same level of removal. Economics of scale were incorporated in the
estimates.

Costs for reducing the discharge of BOD_5 and TSS in the five-commune
area were then estimated as follows. As noted previously, sewer construction areas were added sequentially. For each addition the size of
the collective treatment plant was increased to handle the increased
load.

The sludge generated in the activated sludge process would be dewatered by vacuum filtration to about 20 percent solids content. Most,
if not all, of the resulting sludge would be used by farmers, who would
haul away the sludge, as is currently done at a small treatment plant in
a neighboring town. Thus, the O&M costs of handling and disposing of
dewatered sludge are estimated to be negligible, compared to overall
network and treatment costs. The costs of handling wastewater from
sludge dewatering are included in the treatment costs.

Table 23 shows the capital and O&M costs of increasing degrees of
reduction in BOD_5 and TSS discharges from the five-commune area by means
of adding increments to the collective facility, designed for 95 percent
BOD_5 and 90 percent TSS removal.

Table 23. Estimated Capital and O&M Costs[a] of Reducing BOD_5 and TSS Discharges in a Collective Treatment Plant, Five-Commune Area, 1972 Conditions

	Areas Connected in 1974	1974 + I	1974 + I + II	1974 + I + II + III	1974 + I + II + III + IV	1974 + I + II + III + IV + V
					Connected Areas	
Collection Network						
Capital costs, 10^6 N.D.	None	174	590	764	955	1600
O&M costs, 10^6 N.D./yr.	None	3.4	11.8	15.3	19.1	32.0
Sewage Treatment Plant[b]						
Capital costs, 10^6 N.D.[c]	86	106	115	116	120	123
O&M costs, 10^6 N.D./yr.[d]	3.2	5.9	6.4	6.5	6.7	6.9
Total BOD_5 Generation Collected and Treated (100 metric tons/yr.)	68.6	87.9	101.0	103.0	106.0	108.0
Total BOD_5 Discharged After Treatment (100 metric tons/yr.)	6.7	8.8	10.1	10.3	10.6	10.8
Total TSS Generation Collected and Treated (100 metric tons/yr.)	74.8	80.0	94.6	94.7	97.6	99.6
Total TSS Discharged After Treatment (100 metric tons/yr.)	7.5	8.0	9.5	9.5	9.8	10.0
Sludge Generated[e] (100 metric tons/yr.)	5.2	5.6	6.6	6.6	6.8	7.0

a All costs are rounded and in 1974 N.D.
b 95 percent BOD_5 removal; 90 percent TSS removal.
c Includes capital costs of sludge handling.
d Includes O&M costs for sludge handling and disposal.
e 20 percent solids.

Estimating Costs of BOD_5 and TSS Discharges in On-Site Treatment Plants

Investigation of the locations of industrial plants in the five-commune area indicated that there were twelve plants located outside the area where it was economically feasible to connect to the collective treatment facility. That is, the costs of connecting these plants to the collector system plus the incremental costs of enlarging the collective plant to handle the additional load were larger than the costs of similar reductions in discharges which could be achieved in on-site facilities. Therefore, the cost of reducing BOD_5 and TSS discharges in a treatment plant at each site was estimated.

For each of the twelve industrial plants WW, BOD_5, and TSS generation was estimated, based on its activity subcategory. A primary sedimentation basin was designed for each plant to achieve 28 percent and 73 percent reduction in BOD_5 and TSS discharges, respectively. In addition, for six of the plants it was considered possible to install aerated stabilization basins with secondary sedimentation, which would remove 90 percent of the BOD_5 and TSS input to the basins. Costs were estimated by project personnel and Yugoslav engineers based on actual material and construction cost data. Capital costs include site preparation, construction, equipment, and installation; O&M costs include operating and maintenance labor, materials, energy, and the costs of disposing by landfilling the sludge generated. The capital and O&M costs, and the corresponding reductions in BOD_5 and TSS discharges are shown in table 24, for the installation of these facilities at each of the plants.

Table 24. Estimated Capital and O&M Costs[a] of Reducing BOD_5 and TSS Discharges from Twelve Outlying Industrial Plants in the Five-Commune Area, 1972 Conditions

	Plant Identification Number												Totals
	1	2	3	4	5	6	7	8	9	10	11	12	
Total 1972 Generated & Discharged													
BOD_5, Metric Tons	54.0	0.3	0.1	19.2	42.0	4380	4.7	0.8	2.5	1.1	0.2	0.2	4505
TSS, Metric Tons	64.9	0.4	15.6	23.0	50.2	733	3.6	0.6	2.5	1.1	0.2	25.0	920
Primary Sedimentation[b]													
Capital Costs, 10^6 N.D.	2.1	0.1	0.2	1.1	2.1	11.2	0.3	0.2	0.2	0.1	Neg.	0.3	17.9
O&M Costs, 10^6 N.D./yr.[c]	0.01	Neg.	0.02	0.11	0.13	0.2	0.04	0.01	0.01	Neg.	Neg.	0.05	0.58
Discharge after Primary Sedimentation													
BOD_5, Metric Tons	38.9	0.2	0.1	13.8	30.2	327.0	3.4	0.6	1.8	0.8	0.1	0.1	417
TSS, Metric Tons	18.2	0.1	4.4	6.4	14.1	205.0	1.0	0.2	0.7	0.3	0.1	7.0	258
Aerated Stabilization Plant[d]													
Capital Costs, 10^6 N.D.[c]	2.8	N.A.	N.A.	1.7	3.3	18.6	0.6	N.A.	0.4	N.A.	N.A.	N.A.	27.4
O&M Costs, 10^6 N.D./yr.[c]	0.04	N.A.	N.A.	0.14	0.19	0.57	0.05	N.A.	0.02	N.A.	N.A.	N.A.	1.0
Discharge after Addition of Aerated Stabilization Plant													
BOD_5, Metric Tons	3.9	0.2	0.1	1.4	3.0	32.7	0.3	0.6	0.2	0.8	0.1	0.1	43.4
TSS, Metric Tons	1.8	0.1	4.4	0.6	1.4	20.5	0.1	0.2	0.1	0.3	0.1	7.0	36.6

Abbreviation: N.A., option not applicable; Neg., negligible.

a All costs are rounded and in 1974 N.D.
b 28 percent BOD_5 removal, 72 percent TSS removal.
c Includes O&M costs for sludge handling and disposal.
d Includes secondary sedimentation with 90 percent BOD_5 and 90 percent TSS removal from incoming loads.

Estimating Costs of Handling and Disposing
of Mixed Solid Residuals (MSR)

There are several alternatives for the collection, handling, and
modification of MSR, such as sanitary landfill, materials recovery-
sanitary landfill, shredding, grinding, or baling and landfill, incineration with landfill of incinerator residue, and shredding and combustion
to generate steam or electric energy with landfill of ash. In Yugoslavia
however, only open dumping and low quality landfill operations are presently used. Municipal agencies have only begun to explore the possibilities of using methods of processing MSR other than landfilling.

For the Ljubljana study, it was not possible to obtain sufficient
information from foreign suppliers, as it was for liquid residuals modification, to develop cost estimates for alternative methods of handling
and modifying MSR. Consequently, with the assistance of the municipal
MSR collection agency, capital and O&M costs for a complete sanitary
landfill for all MSR generated in the area were estimated. These costs
were based on the following conditions.

(1) Sufficient land has already been acquired adjacent to the
     present, nearly full, landfill site; thus, no additional
     cost must be incurred for land acquisition. The oppor-
     tunity cost of using this land for a landfill is essen-
     tially zero.

(2) The marsh-like character of this site requires the con-
     struction of a drainage and runoff collection and treat-
     ment system to prevent infiltration of leachate into the
     ground water system, and runoff into the Ljubljanica River.

(3) Several other small landfill sites operating in the five-commune area, which received less than 15 percent of the regional MSR generated in 1974, would be phased out in the near future. However, the location of the proposed site would result in negligible changes in transportation costs to the central landfill site.

(4) No additional collection and transport costs, over and above present costs, will be incurred.

Capital costs for development of the landfill operation include those for: site preparation; construction of a drainage/collection system for surface runoff and infiltrated water; a treatment facility consisting of a primary sedimentation basin and an aerated stabilization basin; and a compactor for compacting the collected MSR. O&M costs include those for: maintenance of the drainage system; operation and maintenance of the wastewater treatment plant; and incremental operation and maintenance of loading and compacting equipment above those currently incurred. Table 25 shows the capital and O&M costs for handling and disposing of the approximately 170×10^3 metric tons of MSR generated annually in the five-commune area.

Table 25. Estimated Capital and O&M Costs[a] of Central Sanitary
Landfill Operation for the Five-Commune Area, 1972 Conditions

	Capital Cost, 10^6 N.D.	Annual O&M Cost, 10^6 N.D.
Drainage System	3.8	0.1
Wastewater Treatment Facility	7.5	0.4
Loading and Compacting Equipment	2.5	0.7
Total	13.8	1.2

a All costs are rounded and in 1974 N.D.

Calculating the Present Value of Costs

The different physical measures for modifying and disposing of residuals have different time streams of costs. Therefore, in order to compare alternatives, costs had to be converted to a common basis. The present value of the time stream of capital and O&M costs, including replacement costs, for the forty-year period of analysis was the basis used. All costs are net costs, and the costs include the capital and O&M costs of handling and disposing of the secondary residuals generated in the modification of primary residuals, such as disposal of fly ash collected in cyclonic filters. Each physical measure has a physical service life ranging from six to forty years, depending on the characteristics of the material items included, at the end of which the salvage value of material items is assumed to be zero. The same physical measure can have different service lives for different activity subcategories.

Economic analysis for decision making for REQM, as for all investment decisions, is complicated by inflation. As long as inflation is
general, e.g., the same for all sectors of the economy, the evaluation
of alternatives is not affected; their relative ranking will be un-
changed. However, even if differential rates of inflation were to occur,
it is virtually impossible to predict those rates, especially in a rapidly
evolving economy such as that of Yugoslavia. Inflation is also one of
several factors which affects the choice of the social rate of discount
to use in discounting the time streams of costs to their present values.
A rate of 11 percent was used, as recommended by Yugoslav experts of
Ljubljanska-Banka. An example of one time stream of costs is shown in
figure 16 for the physical measure of substituting light liquid fuel for
coal in the chemical industry subcategory. As indicated in the figure,
O&M costs increase over time, as the equipment ages, in order to achieve
the same degree of reduction in residuals discharges.

Having estimated the present value of costs for the various physical
measures for the different activity subcategories, the next steps in the
REQM analysis were: (1) to estimate the impacts on AEQ for those residuals for which a natural systems model had been developed; and (2)
to develop and analyze the costs and effects of alternative combinations
of physical measures in relation to the E.Q. targets. These steps are
discussed in chapters VII and VIII, respectively.

Figure 16. Time Stream of Costs for Substituting Light Liquid Fuel for Coal for the Chemical Industry Subcategory, 1972 Conditions

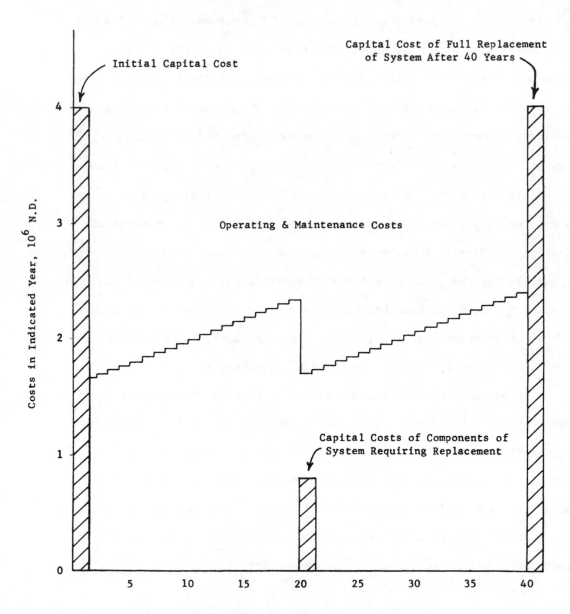

Years After Initiation of Physical Measure

Chapter VII

NATURAL SYSTEMS MODELS

Introduction

Natural systems models estimate the impacts on AEQ of residuals discharged into the environment from man's activities, of course taking into
consideration natural sources of the same materials and/or energy. Thus,
natural systems models must be structured to "accept" as inputs the outputs of activity models, in terms of quantities of specific types of residuals discharged at specific times from specific locations or from
specific areas. Activity models in turn must be formulated in relation
to the natural systems models. If a natural systems model cannot accom-
modate a specific residual in order to determine the quantitative impacts on AEQ, then it is useless to include the residual in the activity
models for the region under study.

As awareness of the adverse consequences of deterioration in AEQ has
grown, increasing attention has been given to the development of natural
systems models to estimate changes in AEQ resulting from residuals discharges. Because of the large existing literature[35] on natural systems
models, the discussion herein is limited to what was done in the Ljubljana
study. However, before turning to that discussion, attention should be
drawn to the following critical issues in the application of natural
systems models in REQM analyses: (a) the relationship between model

---

[35]For example see R.A. Deininger, ed., Models for Environmental
Pollution Control (Ann Arbor: Ann Arbor Science Publishers, Inc., 1973),
and C.S. Russell, ed., Ecological Modeling in a Resource Management Framework, RFF Working Paper QE-1 (Washington, D.C.: Resources for the Future,
1975).

complexity and model accuracy, where accuracy is determined by the close-
ness to which the observed AEQ is predicted, using one or more indicators
of AEQ; (b) the relationship between model complexity and the amount of
information generated; (c) the relationship between model complexity and
model cost; (d) the relationship between amount of information gen-
erated--both level of detail and scope--and the questions being addressed;
(e) the relative accuracy of natural systems models and activity models;
and (f) the relationship between model complexity and the degree to which
observed data are available for model calibration and verification.

Model complexity is a function of the number of variables explicitly
included in the model structure and the effort expended to estimate the
numbers relating to each of those variables. Presumably the accuracy with
which a natural systems model predicts the value or values of one or more
indicators of AEQ resulting from discharges of residuals increases as
model complexity increases. But the rate of increase in accuracy dimin-
ishes with increasing complexity and the cost of modeling increases, as
depicted in figure 17. Accuracy may actually decrease, if the model becomes so complex it is unstable.

Increasing model complexity often increases the amount of information produced. For example, a Streeter-Phelps dissolved oxygen (D.O.)
model yields only estimates of D.O. and BOD_5 concentrations, whereas an
aquatic ecosystem model might yield, for example, not only D.O. and BOD_5
concentrations, but also algae concentration, biomass of fish, and turbidity.

Figure 17. Relationship Among Model Complexity, Model
Accuracy, and Model Cost

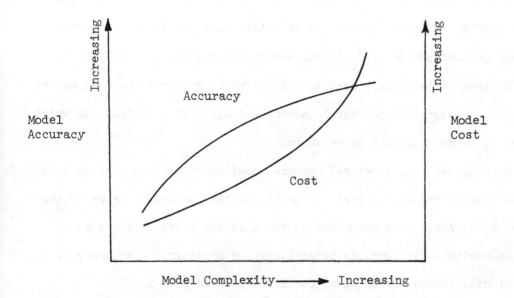

The complexity of a natural systems model to be used in a given case depends on the question(s) being addressed. If the issue of concern is the impact on a particular species of fish, then a simple D.O. model is insufficient. On the other hand, if D.O. is an adequate measure of water quality in relation to recreational use, then the simple D.O. model would be sufficient. If the objective is to obtain an approximate estimate of the mean concentration of SO_2 in the metropolitan area, a box model may be sufficient rather than a diffusion model which requires the delineation of specific residuals discharge locations and specific locations for which ambient concentrations are to be estimated.

The complexity of a natural systems model to be used in a given case depends also on the accuracy of the model relative to the accuracy of the residuals discharge data generated by the activity models. There is no point in having a very complex natural systems model if the accuracy of the residuals inputs generated by the activity models is low.[36]

Finally, if empirical data to calibrate and verify a natural systems model are limited, there is little point in developing a complex model. In such cases there is no way to determine whether or not a complex model is a better predictor of impacts on AEQ than a simple model.

---

[36] See I.C. James II, B.T. Bower, and N.C. Matalas, "Relative Importance of Variables in Water Resources Planning," Water Resources Research vol. 5, no. 6, 1969, pp. 1165-1173.

Raising these issues indicates the difficulty of developing appropriate
natural systems models for any given study. Although there presently
are no definitive rules for determining the degree of model complexity
to be used in a given case, these issues must be considered explicitly.
And they must be considered in relation to the analytical resources available for the natural systems modeling.

In the Ljubljana study, the following decisions were made with respect to natural systems modeling.

(1) Because: (a) deteriorating air quality was considered to be the
most serious AEQ problem in the area; (b) little analysis of air quality
problems had been made; and (c) no empirical data were available with
respect to discharges of gaseous residuals and limited data were available with respect to ambient air quality, it was concluded that relatively
simple modeling of air quality would be adequate and would provide useful
information for present REQM decisions.

(2) Because the lack of water quality data precluded even simple
"first-cut" modeling of ambient water quality, an attempt would be made
only to delineate the dimensions of the water quality problem and to
discuss the implications for water quality modeling with respect to the
five-commune area.

(3) No other natural systems modeling would be undertaken.

This chapter consists of two parts. The first provides a discussion
of the air quality model developed for the five-commune area and the results of its application. The second briefly describes water quality
problems in the five-commune area and their implications for water
quality modeling.

Air Quality Modeling

The objective in analyzing ambient air quality was to construct a
model which would enable evaluation of various physical measures for improving air quality in terms of their costs and impacts on AEQ. The
first step in selecting the type of model to use was to examine the physical and meteorological characteristics of the area. Based on the estimates of total quantities of gaseous residuals discharged, the five-
commune area clearly included most of the air quality problems in the
Ljubljana Basin defined in chapter III. Approximately 90 percent of all
SO_2 and particulate discharges in the basin presently occur within the
five-commune area. Discharges of gaseous residuals from sources outside
the five-commune area do not contribute significantly to ground level
concentrations of SO_2 and TSP within the five-commune area. The five-
commune area also includes 95 percent of the population and approximately
90 percent of the land area in the basin.

Examination of the meteorological records for the eighteen stations in the
five-commune area showed relatively small differences in mean meteorological conditions, such as wind speed, wind direction, and precipitation. The locations of the

eighteen stations are shown in figure 18. Because relatively complete
mixing occurs in the basin, it was assumed that sufficiently accurate
mean daily ground level concentrations of SO_2 and TSP for REQM decisions
are represented by the arithmetic means of the daily measurements at the
eighteen stations.

Based on the above, on the fact that virtually all data with respect to gaseous residuals discharges had to be estimated, and on the
specific questions being investigated, a relatively simple air quality
model--multiple linear regression--was formulated and applied to the
region. Such a model does not <u>explicitly</u> differentiate among locations
of residuals discharges in a region, but--depending on the data--some
inferences are possible about the effects of location on ambient air
quality.

The Air Quality Model

The multiple linear regression model (MLRM) selected is of the
following form:

$$Y_S \text{ or } Y_{TSP} = B_1 + B_2X_1 + B_3X_2 + \ldots B_kX_n + E,$$

where, Y_S or Y_{TSP} = mean ambient concentration of SO_2 and TSP,
respectively, at ground level;

$X_1 \ldots X_n$ = a set of meteorological and residuals discharge
variables;

$B_1 \ldots B_k$ = a set of coefficients to be estimated; and

E = random disturbances.

Figure 18.   Locations of Air Quality Monitoring Stations in Ljubljana Area

Monitoring Stations: Key:

1   Postaja LM Presernova cesta   10   ZD Prule   ———————   Commune Boundary
2   Gospodarska ZB Titova cesta   11   Rozna Dolina
3   HMZ SRS Resljsva cesta   12   Stozice   ══════   Main Road
4   Moste Partizanska ulica   13   Teslova 30
5   Vizmarje Medenska ulica   14   Moste Toplarniska   ///////   Urbanized
6   Avtotehna Titova cesta   15   Krekov Trg   Area
7   ZVD Korytkova ulica   16   Poljanska cesta
8   ZZV SRS Trubarjeva cesta   17   Siska Grosljeva ulica
9   Hala Tivoli cesta   18   Bezigrad Celjska ulica

Having selected the model to be used, the next step was to decide to what periods of the year to apply the model, given the available data and the objective of developing REQM strategies which would be useful in achieving improved air quality. It may be desirable to apply different strategies during some periods of the year than during other periods, if there are significant differences in, and factors affecting, air quality by time of year. Mean monthly concentrations of SO_2 and TSP were computed for each month in the three years, 1970, 1971, 1972, for which years daily measurements of these two indicators were available. The results are shown in table 26. Based on these results, three climatic periods were specified: winter, spring/fall, and summer. The means for the three seasons were sufficiently different so the MLRM was applied separately to each of the three periods.[37]

In order to have sufficient observations for a statistically meaningful sample, the twelve months of daily SO_2 and TSP concentrations for each of the three climatic periods were divided into seven-day periods, and the means for those periods calculated. This yielded 52 observations of the two dependent variables for each of the climatic periods.

The independent variables used in the regression analysis are comprised of two types: (a) meteorological variables; and (b) residuals

---

[37]There are short range air quality problems, e.g., five successive days of extremely poor air quality in a given month, which become evened out in the MLRM.

Table 26. Mean[a] Monthly Concentrations of SO_2 and TSP, Five-Commune Area, 1970, 1971, 1972[b]
(All values in $\mu g/m^3$)

	SO_2			TSP		
	1970	1971	1972	1970	1971	1972
Climatic Period I (Winter)						
November	250	220	240	120	140	130
December	370	490	300	160	250	130
January	260	560	340	170	220	110
February	380	280	280	170	130	130
Winter Mean	315	390	290	130	185	125
Climatic Period II (Spring/Fall)						
March	280	190	210	120	120	100
April	150	100	120	70	50	60
September	80	80	75	60	50	60
October	145	125	125	100	90	80
Spring/Fall Mean	165	125	130	90	80	75
Climatic Period III (Summer)						
May	85	70	110	40	30	50
June	85	55	95	40	30	50
July	85	65	85	60	40	50
August	85	80	70	40	40	50
Summer Mean	85	70	90	45	35	50

a   The mean for a given month is based on means for the eighteen stations at which once a day measurements are made. All values are rounded.
b $\mu g/m^3$ = micrograms per cubic meter.

discharge variables. The values for both are the means for seven-day

periods. The four meteorological variables are:

X_1 -- precipitation per day in mm;

X_2 -- average wind speed per day in m/sec;

X_3 -- meteorological stability per day, five
ordinate classifications:

1. markedly cyclonic;

2. cyclonic;

3. undefinable;

4. anticyclonic;

5. markedly anticyclonic; and

X_4 -- duration of inversion per day, four classes:[38]

0. 0 hrs;

1. $0 < X_4 \leq 8$ hrs;

2. $8 \text{ hrs.} < X_4 \leq 16$ hrs;

3. $16 \text{ hrs.} < X_4 \leq 24$ hrs.

Other meteorological variables affect ambient air quality also, but ex-

perience elsewhere has demonstrated the above are always important.

The residuals discharge variables are the fuel usage activity

categories described in chapter V. The eight residuals discharge

variables are:

----

[38]These classes are based upon measured (3 times daily) temperature
differences, wind differences, and observations of fog between the floor
of the basin (300 m) and the highest observation point in the basin (665 m).

X_5 -- chemical industry;

X_6 -- metal processing industry;

X_7 -- pulp and paper industry;

X_8 -- all other industry;

X_9 -- power plants;

X_{10} -- single-flat residences;

X_{11} -- multi-flat residences; and

X_{12} -- all other activities.

For the regression analysis, the residuals discharged in each of the seven-day periods corresponding to the seven-day means of the meteorological variables had to be estimated. The estimation procedure is described in the appendix to this chapter.

The regression coefficients for each climatic period--for SO_2 and TSP separately--are estimated by using a "backward stepwise", least squares regression routine[39] with a principal components subroutine, to regress the twelve independent variables described above on the dependent variable Y. First, the principal components subroutine is used to form artificial variables. This is done to estimate the partial effects of individual independent variables on the dependent variable, because of multicollinearity between or among variables which are interrelated. These artificial variables are then regressed on Y using the backward stepwise regression routine. The program continues through iterations of

---

[39]The routine is the "Statjob" statistical package developed at University of Wisconsin.

the regression, removing the least significant artificial variable each time, until all remaining variables are significant at the 10 percent significance level, or less, using Student's T test. The remaining artificial variables are then translated to give the explanatory power of the original twelve variables.

The manner in which the principal-component-parts method is used in this study is somewhat unique. In most uses of principal-component-parts in regression analysis it is the equation of regressed artificial variables which is analyzed, not the equation formed by translating the artificial variables back into the original independent variables. Inferences concerning the explanatory power of the original variables are usually drawn by analyzing the significant artificial variables in the regression equation. Because the purpose of developing an air quality model was to predict quantitative changes in ambient SO_2 or TSP concentration resulting from changes in the residuals discharge variables, the use of principal-component-parts was taken one step further than normal.

The manner in which principal-component-parts was used, namely, translating the artificial variables back into the original variables, was verified using SO_2 data for the winter climatic period. It was hypothesized that if the equation formed by translating the artificial variables back into the original independent variables could estimate with reasonable accuracy the standardized mean of the observed SO_2 data by substituting standardized mean values for all of the twelve original independent variables for the same period, then the use of the principal components analysis as described in the text could be considered valid. Substituting the data relating to SO_2 for the winter period and calculating the dependent variable showed that the calculated value of the dependent variable was within about 2 percent of the observed value. This is believed to be sufficient verification of the technique as used in this study.

The MLRM was applied to yield separate relationships for SO_2 and TSP for each of the three climatic periods, a total of six equations. However, only the two equations for the winter climatic period were used for analyzing physical measures to improve ambient air quality, because only during this period does the ambient SO_2 concentration consistently exceed the ambient SO_2 standard of 150 $\mu g/m^3$. The mean ambient concentration of TSP during the winter season is the highest of the three seasons, although the winter mean for the three-year period was slightly less than the ambient TSP standard of 150 $\mu g/m^3$.

The model is used by inserting in the equation changes in the residuals discharge variables that would result from installation of physical measures to reduce discharges, and then calculating the ambient concentration that would result. By repeating the process, the costs of achieving various levels of ambient air quality can be estimated.

Results of Model Application

Table 27 shows the coefficients for the six equations--one each for SO_2 and TSP for the three climatic periods--and the results of the statistical tests applied to the principal components regression equations.

Table 27. Results of Application of the MLRM for Sulfur Dioxide and Particulates[a]

	SO$_2$			TSP		
	Winter	Spring/Fall	Summer	Winter	Spring/Fall	Summer
Statistic Relating to Regressions of Artificial Variables[b]						
Number of significant artificial variables based on Student's T Test, significance level ≤ 0.1	6	4	5	5	4	5
R^2	0.73	0.72	0.54	0.61	0.70	0.65
F Test, significance level ≈ 0.00	24.5	29.8	10.8	14.4	26.9	17.0
Durbin-Watson Test, significance level, 1%	No A.C.[c]	Positive A.C.[c]	Inconclusive	Inconclusive	Inconclusive	Inconclusive
% of Y variance explained in regression	78%	76%	40%	89%	73%	45%
Values of Translated Coefficients of Original Independent Variables (X)						
Meteorological variables:						
X$_1$, precipitation;	-0.10	+0.04	-0.06	-0.00[c]	-0.02	+0.03
X$_2$, average wind speed;	-0.24	-0.15	-0.22	-0.31	-0.14	-0.42
X$_3$, meteorological stability;	-0.13	+0.00[c]	+0.27	+0.11	-0.12	+0.36
X$_4$, duration of inversion.	+0.29	+0.08	+0.09	+0.37	+0.34	+0.02
Residuals discharge variables:						
X$_5$, chemical industry;	+0.17	+0.20	-0.13	+0.13	+0.00[c]	+0.49
X$_6$, metal processing industry;	+0.07	+0.14	+0.21	+0.09	-0.03	+0.10
X$_7$, pulp & paper industry;	+0.07	+0.09	-0.48	+0.17	-0.08	-0.26
X$_8$, all other industry;	-0.10	-0.01	+0.46	+0.02	+0.32	+0.44
X$_9$, power plants;	-0.11	-0.01	-0.11	+0.00[c]	+0.07	-0.22
X$_{10}$, single-flat residences;	+0.19	+0.20	-0.06	+0.05	+0.19	-0.11
X$_{11}$, multi-flat residences;	+0.18	+0.20	-0.07	+0.06	+0.20	-0.04
X$_{12}$, all other activities;	+0.20	+0.20	-0.07	+0.05	+0.2114	-0.04
Constant	+0.32	+0.15	+0.08	+0.15	+0.07	+0.04

Abbreviation: A.C., autocorrelation

a All values are rounded.
b 12 artificial variables were used.
c Slightly greater than zero.

The statistical tests used were: coefficient of determination for the equation (R^2); F test for the whole equation; the Durbin-Watson statistic for autocorrelation; and "Student's T" test for the significance of individual coefficients. On the basis of these tests, sufficient confidence can be placed in the results of the winter period regression equations for SO_2 and TSP so that these equations can be used to estimate the effects of different physical measures to improve ambient air quality. But it should be emphasized that no statistical analysis is meaningful unless there is a physical relationship which is consistent with the statistical results. Careful interpretation is an essential ingredient in the use of statistical models. Including interpretations of the equations derived for each climatic period is deemed useful to provide a fuller understanding of the factors which affect ambient air quality, and to provide a basis for additional analysis of REQM strategies to achieve higher levels of ambient air quality, if and when such levels are desired.

Winter SO_2. For the winter period, the equation shows that precipitation and wind have strong negative effects on ambient SO_2 concentration. Meteorological stability also has a relatively strong negative effect on ambient SO_2 concentration. However, for TSP during the winter climatic period the meteorological stability variable has a positive sign. This inconsistency throws some doubt on the usefulness of this variable, because conceptually the same sign should exist for both SO_2 and TSP. During the summer period meteorological stability is strongly positive for both SO_2 and TSP. Possible explanations for the difference in signs

for the meteorological stability variable between SO_2 and TSP in the

winter are because meteorological stability affects SO_2 and TSP differ-

ently, or because the factor analysis did not sufficiently eliminate the

collinearity between stability and other meteorological variables, or

because of both. The duration of inversion variable was found to have

the highest positive effect on SO_2 concentration. The relative lack of

ventilation during inversion conditions tends to trap SO_2 discharges and

create buildups of SO_2. This is very noticeable in the basin during the

winter period when inversion periods tend to last five to seven days.

In interpreting the explanatory power of the residuals discharge

variables with respect to ambient SO_2 concentration, little checking by

common sense can be done. The coefficients of these variables can only

be compared with limited residuals discharge data, and with the explana-

tory power of the meteorological variables, to gain some sense of the

validity of their relative explanatory powers. The largest effects during

the winter period are from all other activities, and then from single-

flat residences and multi-flat residences, the latter two of which produce

large discharges from low stacks as a result of space heating. The

chemical industry, metal processing industry, and pulp and paper industry,

in that order, exhibit the next strongest effects. However, when all

industrial activities are considered together, they have the largest

effect.

The small negative coefficient for all other industry does not indicate that discharges from those activities improve air quality, e.g., decrease ambient SO_2 concentration. Rather, it indicates that those discharges do not contribute significantly to ambient SO_2 concentrations relative to other dischargers. SO_2 is affected so strongly by other sources of SO_2 discharges that the value for the "all other industry" variable may, in fact, decrease while SO_2 increases, or the reverse, which yields a negative coefficient in the equation. This variable must remain in the equation to maintain a complete set of residuals discharge variables, but it has no significance in connection with estimating future levels of ambient air quality resulting from applying physical measures.

The negative sign on the power plants variable can be readily explained. During the winter season the inversion layer is relatively low.[40] The heights of power plant stacks are such that discharges are above the inversion layer where they are dispersed over a much larger area than they would be if the stacks discharged below the inversion layer. As a result, they do not contribute much to ground level concentrations in the five-commune area. Because very few, if any, stacks from other activities penetrate the inversion layer, stack height does not explain the negative sign for all other industries.

---

[40]This is indicated by visual observations and some limited data and photographs from weather balloons from Hidrometeoroloski Zavod SRS.

Summer SO_2. Conditions during the summer period are not as uniform as during the winter period. Therefore, physical interpretation of the equation for this period is more difficult. Precipitation and wind have negative effects on SO_2 concentration. However, because summer is usually the driest season the negative coefficient for precipitation is small. The meteorological stability variable is strongly positive for both SO_2 and TSP, which supports the hypothesis that regional anticyclonic conditions during the summer season tend to induce inversion conditions and result in the build up of ground level SO_2 concentration. The duration of inversion variable is positive but small, probably due to the pattern of short daily inversion periods experienced during early morning hours in the summer season.[41]

Only two residuals discharge variables have any positive effect on ambient SO_2 concentration in the summer period: metal processing industry and all other industry. This reflects two facts. First, space heating in industrial operations--and hence residuals discharges therefrom--are negligible during the summer. Second, many industrial operations have seasonal patterns of production, with lower levels of operation in the summer period. Thus, residuals discharges originating from production processes are substantially less during the summer than during

---

[41]During the summer season cooler air settling in the basin during the evening is trapped, causing periods of inversion during early morning hours until solar radiation warms the cold air masses.

the other seasons. However, the "all other industry" variable represents primarily food processing activities which have their high period of production during the summer; this is the likely explanation for its large positive coefficient.

Discharges from power plants do not decrease significantly during the summer, but their impacts are still affected by stack heights being above the low inversion layers, as in winter. The negative signs for single-flat residences, multi-flat residences, and all other activities are explained by the fact that these activities have minimal discharges during the summer, stemming primarily from hot water heating; space heating is negligible.

Spring/Fall SO_2. During this season daily precipitation has a very small but positive effect on SO_2 concentration. This may be the result of either inaccurate data or the times at which rain occurs or both. Rainfall during the night may not have much effect on daytime ambient concentration resulting from industrial or residential discharges. The effects of the meteorological stability and duration of inversion variables are relatively small and positive, explaining little of the variation in SO_2 concentration. Chemical industry, multi-flat residences, all other activities, single-flat residences, and metal processing industry all have strong positive explanatory power decreasing in that order. Again, all other industries and power plants have negative signs, probably for the same reasons as for the winter season.

Winter TSP. During the winter period, wind and duration of inversion affect TSP concentration much as they did SO_2 concentration, having strong negative and positive effects, respectively. However, precipitation, although affecting TSP negatively, does not seem to have nearly as strong an effect on TSP as on SO_2. One possible explanation is the relatively stronger effect wind has on TSP as compared to its effect on SO_2. This larger effect is the result of the larger particle size and physical characteristics of suspended particulates. Meteorological stability has a relatively strong positive effect on TSP, but the validity of this result is suspect, as suggested above.

Of the residuals discharge variables, chemical industry and pulp and paper industry have the greatest effects on TSP concentrations. Single-flat residences, multi-flat residences, and all other activities are the next strongest variables, in that order. Single-flat and multi-flat residences are not the strongest variables, as they are for winter SO_2, because these activities tend to burn a greater proportion of liquid fuels than industrial activities. The liquid fuels result in significant SO_2 discharges but small particulate discharges. The coefficients for all other activities and power plants are both positive but very small, which is not conclusive in explaining variations in TSP. These results may be partly due to the same reasons cited in interpreting the equation for SO_2 during the winter period, and partly due to the slightly different physical behavior of particulates compared to SO_2.

Summer TSP. No conclusions can be drawn concerning the effect of precipitation on TSP during the summer because of the small positive coefficient. However, wind again has a strong negative effect. Meteorological stability indicates that anticyclonic conditions correlate positively with TSP concentrations. The duration of inversion variable again correlates positively with TSP; however, the coefficient is small, probably due to the very short daily inversion periods during the summer.

Only three residuals discharge variables have positive effects on TSP. Two of these were the only positive variables in the summer SO_2 equation, metal processing industry and all other industry. That chemical industry also has a high positive coefficient may result from the fact that the chemical industry uses a higher proportion of solid fuels-- which result in greater quantities of particulate discharges than liquid fuels--than do other industrial activities. The remaining variables all have negative coefficients, probably for the same reasons cited for summer SO_2.

Spring/Fall TSP. Precipitation, wind, and duration of inversion all are negatively correlated with TSP concentration. Meteorological stability indicates that stable anticyclonic conditions correlate negatively with TSP. Two residuals discharge variables have negative signs, the metal processing industry and pulp and paper industry. All the rest are positive, with all other industry having the greatest effect.

The validity of the above interpretations can be questioned, but there is no way fully to confirm or disprove them. Credibility of the analysis and interpretation lies in the validity of the assumptions used in applying the statistical model, the quality of the data used, the method by which the output has been extracted (using acceptable statistical procedures and tests), and the ability to offer physical explanations for results. These results must first be considered acceptable before one can use the statistical relationship to estimate the effects of various measures to reduce discharges on ambient SO_2 and TSP concentrations, and hence to evaluate strategies for the improvement of ambient air quality.

One additional point merits mention. Although a statistical model such as the MLRM used in the Ljubljana study does not explicitly consider the locations of the discharges, it does so implicitly. That is, discharges located closer to the points at which ambient concentrations are measured will have greater impact on--and hence show higher correlations with--the measured ambient concentrations than discharges located farther away, all other factors being equal.

Water Quality Modeling

The decision not to model water quality in the five-commune area was a result of limited analytical resources and two other considerations. First, the five-commune area comprises only part of the Upper Sava River Basin, as shown in figure 7 (chapter III). Discharges upstream from Ljubljana can--and apparently do--affect water quality in the five-commune area, particularly surface water quality. Therefore, the development of efficient REQM strategies to improve ambient water quality in the five-commune area would require analysis of the entire Upper Sava

River Basin. Second, data for assessing the impacts of alternative
physical measures for managing surface and ground waters were lacking,
including variation in present ambient water quality by time of year,
infiltration/recharge capacity, and aquifer yields.

Nevertheless, placing the area's water quality problems in perspec-
tive in the context of REQM and suggesting relevant natural systems
modeling for generating information useful for REQM decisions were important components of the REQM analysis. Demands on the surface and
ground water resources of the area will increase over time, with associated likely decreases in quality of those resources unless counter-
measures are undertaken. For example, plans exist for: the recharge of
Ljubljana well fields with approximately ten liters per second from the
Sava River, which quantity represents less than 3 percent of the Sava's
minimum flow but almost 15 percent of total water pumped; the filling
of wetlands in the five-commune area; and the construction of sewage
treatment plants. Assessing the impacts of these plans and of the increased demands for water requires quantitative modeling, if rational
water quality management is to be achieved in the area. Water quality
models for these purposes can be suggested. Doing so has the additional
utility of defining what data need to be collected to supplement the
limited existing data to enable at least "rough-cut" water quality modeling to be undertaken in the near future. As a partial step in this

direction, a related project study was undertaken which focused on management of surface water quality in the Upper Sava River Basin.[42] The
following discussion of water quality problems in the five-commune area
and the implications of those problems for water quality modeling are
based primarily on that study.

Water Quality Problems in the Five-Commune Area

The Sava River Basin includes areas in the republics of Slovenia,
Croatia, and Serbia, and extends to the confluence of the Sava River with
the Danube River at Beograd. The five-commune area is part of the Upper
Sava River Basin, which basin encompasses twenty communes with a 1972
population of 635,000 of a total population of 1,725,000 in Slovenia.
The boundaries of the Upper Sava River Basin are mostly congruent with
the jurisdictions of three middle and Upper Sava water management communities.[43]

---

[42] T.R. Angotti, "Planning and Management of Water Resources in the
Ljubljana Region: Applying the REQM Framework," (Working Paper, The
Johns Hopkins-Urbanisticni Institut Environmental Research Project,
Ljubljana, Yugoslavia, 1975), and T.R. Angotti, "Planning and Manage-
ment of Water Resources in the Ljubljana Region of Yugoslavia,"
Environmental Conservation vol. 3, no. 3, Autumn 1976. Stemming from
the Angotti study a first attempt to model surface water quality in
relation to dissolved oxygen was made. This is reported in L.C. Koss,
"Applied BOD Residuals Environmental Quality Management in the Upper
Sava River Basin," (Working Paper, The Johns Hopkins-Urbanisticni Institut
Environmental Research Project, Ljubljana, Yugoslavia, November 1975).

[43] Water management communities were created under the 1974 National
Constitution, which states: "Self-managing communities of interest
shall be formed by working people, directly or through their self-managing organizations, to satisfy their personal and common needs and interests." These needs and interests include such components of water
management as water supply, flood damage reduction, and water quality
management. Water management communities are organized according to
commune boundaries.

Only about 3 percent of the liquid residuals generated in residential, commercial, institutional, and some industrial activities in the
Upper Sava River Basin receive any degree of modification before discharge. Although there is no complete survey of industrial sources discharging liquid residuals directly into the Upper Sava River or its tributaries, the information available indicates that most of the 150 major
industrial generators of liquid residuals provide no modifications of
those residuals before discharging directly into the surface waters of
the Upper Sava River Basin. Organic materials are discharged from
various food processing industries--especially large discharges from
breweries and from dairy and wine producers--and inorganic materials from
metal and chemical processing activities. Of the five major sources of
BOD_5 and TSS residuals discharged into the Upper Sava River, the two
largest are the Ljubljanica River and a pulp and paper plant near the
Croatian border. The former receives discharges from industrial activities and municipalities most of which are unmodified.

The five-commune area is located approximately midway between the
Sava headwaters and the Croatian border and accounts for about 25 percent of all discharges of BOD_5 and TSS residuals in the Upper Sava
River Basin. Most municipal liquid residuals generated in the five-
commune area are discharged at the single outfall of the Ljubljana sewer
system. The five municipal secondary wastewater treatment plants in the
five-commune area are small and treat only about 2 percent of all collected wastewater. The main outfall of the Ljubljana sewer system is

just downstream from one of two flood barriers across the Ljubljanica
River and upstream from the other. These flood barriers inadvertently
serve, to some extent, as settling ponds.

Approximately 65 percent of the urban population of Ljubljana is
served by public sewers, while 90 percent is connected to the public
water supply system. The Ljubljana sewer system, portions of which were
as much as 78 years old in 1972, is plagued by maintenance and infiltration problems in some areas. A sizeable portion of the system combines
sanitary sewage and storm water runoff. At least two other problems
exist. One is poor maintenance in many of the approximately 100 small
local water supply systems. Another stems from the location and the
type of construction or both of many individual household wells, such
that public health hazards may exist because of infiltration from surface
drainage and/or septic tanks.

The water table in the Ljubljana aquifer, which is the source of
most potable and industrial water, has declined from 2 to 8 meters since
the turn of the century, because of increased withdrawals and subseqeunt
discharge of wastewater into the Sava River downstream from the five-
commune area. Although it appears that existing sources of water are
adequate to meet projected water withdrawals to 1980, additional sources on
more efficient use of water from existing sources or both will likely
be required in the long run. The costs of providing water to users is
exacerbated by losses of water in the distribution system, which losses
are estimated to be as much as 30 percent of the total water pumped.

Six important questions with respect to water resources management in the five-commune area have been identified. First, what is the "safe yield" of the ground water aquifer, considering current trends in magnitudes and locations of water withdrawals? Second, what are the effects of surface water quality and use and land use on the quality and quantity of water in the aquifer? Third, what are the effects of drawdown and/or poor water quality in one portion of the aquifer system on other parts of the same system? Fourth, what are the effects of drawdown and/or poor water quality in the aquifer on surface waters? Fifth, what are the effects of upstream discharges of residuals on surface and ground water quality in the five-commune area? Sixth, what are the effects of residuals discharges in the five-commune area on water quality downstream?

With respect to the first question, the level of the water table in 1972 indicated that about 75 meters remained in the water-bearing stratum of the aquifer. However, the productivity of this stratum has not been clearly determined. With respect to the second question, there have been no comprehensive studies of the effects of land use on urban storm water runoff and agricultural runoff, and in turn of the effects of such runoff on groundwater. However, surveillance of groundwaters and special studies have revealed the presence in ground water of such substances from human activities as mineral compounds and detergents. Traces of phenols were discovered in the Ljubljana aquifer after the recent cleaning of an upstream dam on the Sava River. Similarly, the impacts of approximately 2 million cubic meters per year of septic tank discharges have

also not been ascertained. With regard to the third question, little
research has been done concerning the movement of water within the aquifer;
hence, little is known about effects of drawdown and/or poor water quality
in one segment of the aquifer on quantity and quality in other segments.
Similarly, the effects of drawdown and/or poor water quality in the
aquifer on surface waters, the fourth question above, are also not known
at this time.

With respect to the fifth and sixth questions, figure 19 shows a
dissolved oxygen profile of the Sava River. The profile appears to represent a classical oxygen sag. The low point of the dissolved oxygen
curve results from discharges of organic material in Slovenia--both upstream from, and at, Ljubljana--and in Croatia. Velocities of approximately 2 meters per second will carry organic residuals 180 to 200 kilometers a day. Thus, discharges of organic materials made upstream from
Ljubljana exert a demand for oxygen at Ljubljana as shown, and together
with discharges of organic materials from Ljubljana, still exert a significant
demand for oxygen in the vicinity of Zagreb and Sisak. At present, data
are insufficient to determine what proportion of the oxygen sag is a
result of discharges upstream from Zagreb and what proportion is a result
of discharges at Zagreb. However, it appears that Slovenia's contribution to the dissolved oxygen sag is significant.

Implications for Water Quality Modeling

Mathematically expressed models of river water quality and ground
water quality would contribute to decision making in the five-commune

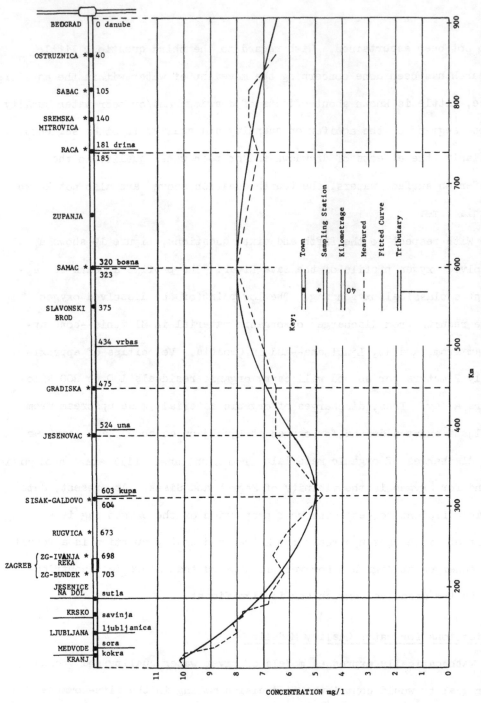

Figure 19. Dissolved Oxygen Profile for the Sava River

Source: Computed based on data taken from: Polytechna-Hydroprojekt-Carlo Lotti and Co., The United Nations Study for the Regulation and Management of the Sava River in Yugoslavia, B/11-Water Utilization Plan, United Nations Publications, Prague-Roma, April, 1972.

area, both by improving knowledge of existing stream and aquifer behavior and by aiding in the prediction of changes in water quality resulting from alterations in stream flow, ground water withdrawals, and discharges of liquid residuals. Such models are essential tools in the evaluation of alternative water supply and water quality management strategies. Based on the foregoing description of water quality and water supply problems in the five-commune area, recommendations can be made concerning the extent and types of water quality modeling that might usefully be undertaken in the future.

First, a water quality model of the Ljubljanica River does not need to be developed at present for management decisions. Because the water quality of the Ljubljanica is so poor--as a result of receiving most of the liquid residuals discharges from the five-commune area from both point and nonpoint sources--little insight into the investment decisions needed to improve water quality would be gained by investing scarce resources to model the Ljubljanica. Rather, efforts should be expended to obtain estimates of the total quantities of organic materials and suspended sediment discharged into the Ljubljanica from different point and non-point sources, and the costs to reduce such discharges. It is likely that such information would indicate where resources should be invested first to begin to improve water quality in the Ljubljanica.

Second, because reducing liquid residual discharges in the five-commune area would have little or no effect on water quality of that segment of the Sava River in the five-commune area, modeling only that

segment of the Sava River would be of little value to decision makers
within the five-commune area. The quality of the Sava River in the five-
commune area can only be affected by reduction in residuals discharges
upstream from the area.[44] If any model of the Sava River is to contribute to selection of water quality management strategies in the five-
commune area, the modeling effort must include at least the entire Upper
Sava River Basin. Given that data on existing water quality and residuals discharges for the Upper Sava River Basin are limited, and that the
first water quality problem to be assessed should probably be limited to
organic materials and dissolved oxygen, a simple Streeter-Phelps type
model should be sufficient, using estimated discharges from both point
and nonpoint sources.

Third, with respect to improving the capacity to manage the aquifer
system, a model of flow into, within, and from the aquifer will be required within the next decade. Among the questions that must be answered
concerning the aquifer are those relating to the effects of the magnitudes
and locations of pumping from the aquifer and/or recharging the aquifer
on the yield of and water quality in the aquifer. Because of the limited
information available concerning the Ljubljana aquifer system, it is not
possible to recommend a specific modeling approach at this time, only
that efforts to develop a model of the aquifer should be given high priority.

---

[44]However, the application of residuals discharge reduction measures
to activities in the five-commune area will have significant impacts on
the quality of the Sava at Zagreb, in Croatia, as shown by figure 19.

Appendix to Chapter VII

ESTIMATING SO$_2$ AND PARTICULATE DISCHARGES FOR USE
IN THE MULTIPLE LINEAR REGRESSION AIR QUALITY MODEL

In order to enable a meaningful statistical analysis with the multiple
linear regression model, daily discharges of SO$_2$ and particulates from
each of the fuel usage categories identified in chapter V were estimated.
Seven-day means were then calculated for use as inputs to the regression
model.

Given the variability in the coefficients used and in the fuel consumption data, it is unrealistic to assume that estimates of daily dis-
charges--or even the seven-day averages used--are anything but approximate
measures of actual discharges. This variability problem exists regardless
of the natural systems model used. One advantage in using a statistical
model, such as a linear regression formulation, is that it is less sensi-
tive to inaccuracies in absolute values because it scales all variables
appearing in the formulation, and yields relative effects among variables
rather than absolute effects.

The method used in estimating discharges for a given period was to
multiply the quantity of each type of fuel used during the period by the
relevant discharge coefficient, kilograms of residual discharged per unit
of fuel burned, given that discharge is the same as generation, except for
the one power plant, as noted previously. It is very difficult to obtain

fuel use data on a daily or even a weekly basis. Therefore, a series of assumptions had to be made to estimate fuel use, and hence discharges, on a daily basis. Depending upon the activity subcategory, the assumptions varied slightly. The method is illustrated by application to industrial activities.

The basic assumption is that SO_2 and particulate discharges from industrial activities result from the combustion of fuels for space heating, for production process purposes, or both. Combustion for heating water is small compared with the other two purposes. Separately, each of these components can usually be allocated without much difficulty: space heating discharges by degree days;[45] production process discharges by some productivity measure, e.g., pieces produced per day. A problem arises when both components occur simultaneously in a given activity. Figure 20 illustrates this problem by showing hypothetical SO_2 discharges from an industrial activity over a one-year period.

---

[45]A degree day is a commonly used term in meteorology and is sometimes referred to as a "heating day." In the Ljubljana area, it is defined as a one-degree declination from $12^{\circ}C$ for a given day. It was estimated that $12^{\circ}C$ is the approximate temperature below which space heating commences. This is substantially different from the $16^{\circ}C$ used in the U.S. Ljubljana's power plants begin to provide space heating after September 15, but only after three consecutive days have had a temperature below $12^{\circ}C$ at 9:00 p.m.. They stop heating after May 15 or whenever three consecutive days before May 15 have had temperatures above $12^{\circ}C$.

Figure 20. Hypothetical Mean Monthly Rates of SO_2 Discharge From an
Industrial Activity over a One-Year Period

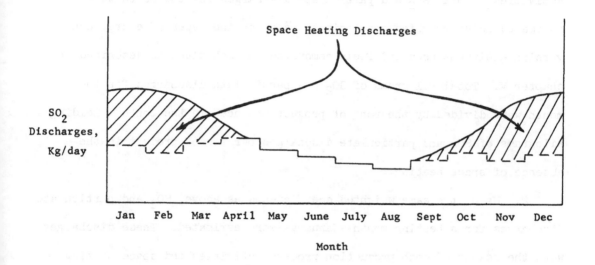

The figure assumes that production, and hence production process discharges, vary some throughout the year. It illustrates that during the summer period discharges from the industrial activity are attributably only to production activities. Discharges from space heating and production processes occur jointly only during the colder space heating months. These facts comprise the basis for the estimation procedure.

Given information on product output and fuel use by month for each industrial plant and daily temperature data for the three-year period, 1970-1972, daily discharges were estimated, and seven-day averages then calculated for each industrial plant. A description of the procedure follows.

1. A summer--non-space heating--month (June) was selected during which gaseous residuals discharges resulted only from production process activities. Total SO_2 and particulate discharges for the month were estimated using weighted coefficients based on the types of energy conversion equipment used and fuels combusted in each plant as described in chapter V. Total kilograms of SO_2 and particulate discharges for the month were divided by the tons of product produced in the month, yielding kilograms of SO_2 and particulate discharges per ton of product in the absence of space heating.

2. Using the same weighted coefficients as above, SO_2 and particulate discharges for a heating month--January--were estimated. These discharges were the result of both production process activities and space heating. The discharges in January attributable only to production processes were estimated by multiplying the kilograms of discharge per ton of product output produced in the absence of space heating, computed in 1 above, by the tons of product produced in the month. The estimated discharges attributable to space heating then were the differences between the total discharges estimated for January and the discharges attributed to production processes.

3. Total SO_2 and particulate discharges in January attributable to space heating were then divided by the total number of degree days in the month, yielding kilograms of discharges from space heating per degree day. This relationship is a function of such factors as facility size, building design, insulation, heat generated by production processes and number of employees, factors which vary from plant to plant.

4. Having separated the space heating and production process components, the estimation of daily SO_2 and particulate discharges was straightforward. Daily discharges from production process activities were estimated by multiplying kilograms of residuals discharged per ton of product output in the absence of space heating times product produced in each month. These total amounts were distributed equally throughout the month, assuming the level of output per day in the month was constant. Daily discharges from space heating activities were estimated by multiplying the previously estimated kilograms of discharge per degree day times the number of degree days in each day. Total daily discharges were then the sum of the daily discharges due to production process activities and the discharges due to space heating activities for each day for the three years for which data were available. Seven-day averages were then computed from these daily data.

The estimated monthly discharges of SO_2 in the five-commune area by major activity categories for 1970, 1971, and 1972 are shown in figure 21.

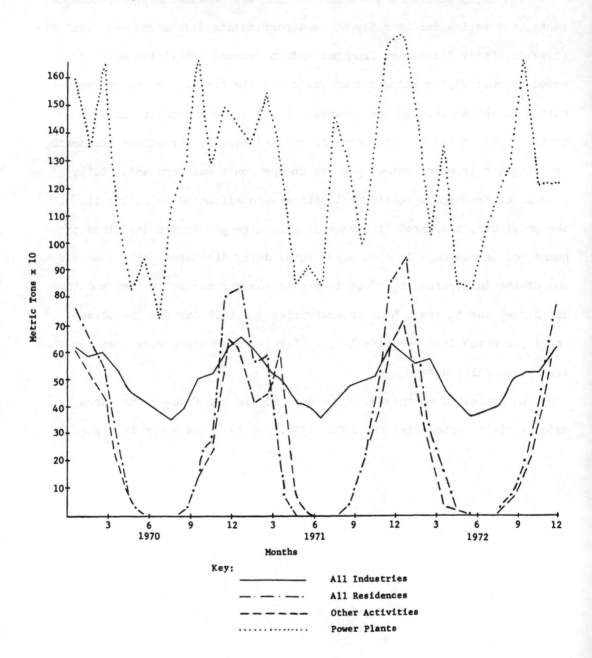

Figure 21. Estimated Monthly Discharges of SO_2 for 1970, 1971 and 1972, by Activity Categories

Metric Tons x 10

Months

1970   1971   1972

Key:

——————————— All Industries

—·—·—·—·— All Residences

— — — — — Other Activities

············· Power Plants

Chapter VIII

DEVELOPING AND ANALYZING COMBINATIONS OF
PHYSICAL MEASURES FOR REDUCING RESIDUALS
DISCHARGES AND IMPROVING AMBIENT ENVIRONMENTAL QUALITY

As defined in chapter II, an REQM strategy consists of: (1) the
physical measures for reducing the generation and/or discharge of residuals to the environment and for directly increasing environmental assimilative capacity; (2) the implementation incentives which induce the
individual activities--households, industrial operations, institutions,
agricultural operations--to adopt the physical measures; and (3) the
institutional arrangement or set of institutions with the authorities to
impose the implementation incentives on the individual activities and to
carry out other functions of REQM. This chapter deals with the first
element of REQM strategies--physical measures--and describes how combinations of physical measures were developed in relation to E.Q. targets and
then analyzed. The second and third elements are discussed in chapter IX.

A combination of physical measures can be comprised of a single
physical measure directed at one residual from a single activity, various
physical measures directed at one residual generated by each of multiple
activities, various physical measures directed at several residuals
generated by one or more activities. The number of combinations of
physical measures developed in a given REQM analysis is a function of:
(1) the number of activities; (2) the number of different physical measures developed for reducing discharges of residuals from each individual
activity; (3) the number of alternative levels of E.Q. investigated; and
(4) the available analytical resources. In the Ljubljana study, a base
set of E.Q. targets was specified, as shown in table 6 in chapter IV.

In order to show the costs of achieving different levels of E.Q. and to illustrate the tradeoffs among residuals and environmental media, four other sets of environmental quality targets were analyzed.

Developing and Analyzing Combinations of Physical Measures

Table 28 summarizes the physical measures analyzed for modification of residuals, the residuals which they are to modify, and the major secondary residuals generated, if any, by application of each measure. These measures, their costs, and the residuals discharge reductions related thereto were described in chapter V. Some of these measures are mutually exclusive, that is, they cannot be applied simultaneously to the same activity. For example, desulfurization of heavy liquid fuel cannot be applied at the same time as shifting from heavy liquid fuel to light liquid fuel. Similarly, the effluent from an on-site primary sedimentation and aerated stabilization plant is not likely to be discharged into a collective activated sludge facility. Each physical measure was applied at only one level of intensity, e.g., cyclonic filters were designed to remove 95 percent of the particulates from the discharge stream. However, some measures could be applied to more or fewer of the individual activities, e.g., off-site central heating.

Effectiveness of Physical Measures

Having calculated the present value of costs for each physical measure applied to each relevant activity subcategory--as discussed in chapter VI--the next step was to calculate the cost effectiveness of each measure

Table 28. Physical Measures for Residuals Modification Analyzed in the Ljubljana REQM Study

REQM Physical Measure	Residual Affected							
	Gaseous					Liquid		Solid
	SO_2	Partic-ulates	CO	HC	NO_x	BOD_5	TSS	MSR and Sludge
Cyclonic filter to remove 95% of particulates in stack gas		X[a]						O[b]
Substitution of light liquid fuel with 1% sulfur, essentially no ash, for coal with 2% sulfur, 12.5% ash	X	X	X	X	X			X[c]
Substitution of light liquid fuel with 1% sulfur, essentially no ash, for heavy liquid fuel with 3% sulfur essentially no ash	X	X	X	X	X			
Desulfurization of heavy liquid fuel from 3% sulfur to 0.5% sulfur[d]	X	X						
Off-site central heating from coal-fired power plants to all activities in feasible connection area except industrial and power plants	X	X	X	X	X			O
Park/Ride, to reduce daily vehicle kilometers traveled in five-commune area			X	X	X			
Carburetor idling adjustment applied twice a year to all gasoline-fueled vehicles in five-commune area			X	X	X[e]			
Primary sedimentation/activated sludge, collective wastewater treatment plant						X	X	O
On-site primary sedimentation/aerated stabilization wastewater treatment plant						X	X	O
Sanitary landfill								X

a X indicates measure applies to indicated residual.

b O indicates application of measure results in generation of secondary residual of type indicated.

c The indicated measure results in a decrease in the quantity of ash to be handled in the five-commune area.

d Desulfurization is assumed to be done outside the five-commune area.

e The indicated measure results in an increase in discharge of NO_x.

applied in relation to the relevant residual, e.g., the amount of environmental quality which could be "purchased" by a given expenditure on a
physical measure-activity subcategory combination. This calculation
was in terms either of: (1) reduction in ambient SO_2 or TSP concentration per 10^6 1974 New Dinars (N.D.) of expenditures; or (2) reduction in
discharge of BOD_5, TSS, HC, NO_x, and CO, per 10^6 1974 N.D. As a measure
of cost effectiveness, either index is limited because the same physical
measure applied to an activity subcategory will often reduce the discharge of more than one residual. For example, the Park/Ride measure
will simultaneously reduce discharges of HC, CO, and NO_x.

Tables 29, 30, 31, and 32 present illustrative data. The first two
show physical results in relation to discharges of SO_2 and particulates,
respectively. Column 1 indicates the physical measure and column 2, the
fuel usage category to which it is applied. Column 3 indicates the
maximum reduction in discharges from the fuel usage category, measured
as the percent of present discharges from the category; column 4, the
maximum reduction in discharges of the residual in the five-commune area,
measured as the percent of present total discharges in the area. Maximum
in both cases means that the physical measure is applied to all individual
plants or units in the fuel usage category at the highest degree of
intensity considered. Column 5 indicates the maximum improvement in
ambient concentration in the winter period, measured as the percent
reduction from the average winter concentration under present conditions,
e.g., about 330 $\mu g/m^3$ for SO_2 and 150 $\mu g/m^3$ for particulates in tables
29 and 30, respectively. Table 31 shows the present value of costs to

Table 29. Effects of Physical Measures to Reduce SO$_2$ Discharges, 1972 Conditions, Five-Commune Area[a]

1	2	3	4	5
Physical Measure	Fuel Usage Category	Maximum reduction in SO$_2$ discharges from activity sub-category, % from base activity discharge	Maximum reduction in SO$_2$ discharges in region, % from base regional discharge	Maximum reduction in mean winter ambient SO$_2$ concentration, % reduction from base[b] concentration
Substitution of light liquid fuel with 1% sulfur, essentially no ash, for coal with 2% sulfur, 12.5% ash	Chemical industry	42	1	7
	Metal processing industry	74	2	6
	Pulp & paper industry	59	5	19
	All other industry	30	2	c
	Multi-flat residences	64	5	11
	Single-flat residences	64	3	11
	All other activities except P.P.	73	8	14
	Total	--	26	68
Desulfurization of heavy liquid fuel from 3% sulfur to 0.5% sulfur	Chemical industry	38	<1	7
	Metal processing industry	6	Negligible	Negligible
	Pulp & paper industry	24	2	8
	All other industry	50	4	c
	Multi-flat residences	8	1	1
	Single-flat residences	8	<1	1
	All other activities except P.P.	6	1	1
	Total	--	9	18
Substitution of light liquid fuel with 1% sulfur, essentially no ash, for heavy liquid fuel with 3% sulfur essentially no ash	Chemical industry	30	Negligible	5
	Metal processing industry	5	0<%<2	Negligible
	Pulp & paper industry	19	3	6
	All other industry	41	<1	c
	Multi-flat residences	6	<1	1
	Single-flat residences	7	<1	1
	All other activities except P.P.	5	<1	1
	Total	--	8	14
Off-site central heating from[d] fossil fuel-fired power plants	All activities except industrial and P.P.	23[e]	-1 (net increase)	c

Abbreviation: P.P., power plants

a All values are rounded.

b Base refers to 1972 conditions.

c The negative coefficient in the multiple linear regression model for air quality for this measure applied to this activity indicates that no significant improvement in SO$_2$ concentration is achieved.

d All five feasible connection areas are connected.

e SO$_2$ discharges from the units in activities switched to off-site central heating are reduced by 23%, but SO$_2$ discharges from power plants are increased by 20%, resulting in a net increase in SO$_2$ discharges of 1%.

Table 30. Effects of Physical Measures to Reduce Particulate Discharges, 1972 Conditions, Five-Commune Area[a]

1	2	3	4	5
Physical Measure	Fuel Usage Category	Maximum reduction, particulate discharge from activity category, % from base[b] activity discharge	Maximum reduction, particulate discharges in region, % from base[b] regional discharge	Maximum reduction in mean winter ambient TSP concentration, % reduction from base concentration
Substitution of light liquid fuel with 1% sulfur, essentially no ash, for coal with 2% sulfur, 12.5% ash	Chemical industry	87	4	8
	Metal processing industry	90	11	10
	Pulp & paper industry	91	18	30
	All other industry	75	9	19
	Multi-flat residences	70	7	4
	Single-flat residences	68	5	3
	All other activities except P.P.	78	10	3
	Total	--	63	77
Cyclonic filter to remove 95% of particulates in stack gas	Chemical industry	86	4	8
	Metal processing industry	83	10	9
	Pulp & paper industry	91	18	30
	All other industry	78	9	20
	Multi-flat residences	79	8	4
	All other activities except P.P.	89	11	7
	Total	--	60	55
Off-site central heating from fossil fuel-fired power plants[d]	All activities except industrial and P.P.	13[e]	5	9

Abbreviation: P.P., power plants

a All values are rounded.
b Base refers to 1972 conditions.
c Cyclonic filters are not applicable to single-flat residences.
d All five feasible connection areas are connected.
e Discharges of particulates from the units in activities switched to off-site central heating are reduced by 13%, but discharges of particulates from power plants are increased by 21%, resulting in net decrease of 5% in the region.

Table 31. Present Value of Costs to Achieve a 1 Percent Reduction in Ambient SO_2 Concentration, 1972 Conditions, Five-Commune Area.

Physical Measure	Fuel Usage Category	Present value of costs, 10^6 N.D., per 1% reduction in ambient SO_2 concentration	Cost Effective Ranking	
			With respect to single measure	With respect to all measure-activity combinations
Substitution of light liquid fuel with 1% sulfur, essentially no ash, for coal with 2% sulfur, 12.5% ash	Chemical industry	375	1	2
	Metal processing industry	1130	4	5
	Pulp & paper industry	750	2	6
	All other industry[b]	--	--	--
	Multi-flat residences	1390	5	8
	Single-flat residences	915	3	7
	All other activities except P.P.	1670	6	16
Desulfurization of heavy liquid fuel from 3% sulfur to 0.5% sulfur	Chemical industry	300	1	1
	Metal processing industry	560	3	11
	Pulp & paper industry	525	2	3
	All other industry[b]	--	--	--
	Multi-flat residences	950	4	13
	Single-flat residences	970	5	9
	All other activities except P.P.	1200	6	12
Substitution of light liquid fuel with 1% sulfur, essentially no ash, for heavy liquid fuel with 3% sulfur, essentially no ash[c]	Chemical industry	540	1	4
	Metal processing industry	1660	4	15
	Pulp & paper industry	1120	2	10
	All other industry[b]	--	--	--
	Multi-flat residences	2240	5	17
	Single-flat residences	1550	3	14
	All other activities except P.P.	2370	6	18
Off-site central heating from fossil fuel-fired power plants[d]	All activities except industrial and P.P.[b]	--	--	--

Abbreviation: P.P., power plants

a   Costs are in 1974 N.D.; 40-year period of analysis; 11% discount rate.
b   Application of measure to activity results in no significant reduction in ambient SO_2 concentration.
c   Measure cannot be implemented simultaneously with desulfurization of heavy liquid fuel.
d   All five feasible connection areas are connected. SO_2 discharges from units connected are reduced, but SO_2 discharges from the power plants are increased, resulting in no significant reduction in ambient SO_2 concentration.

Table 32. Present Value of Costs to Achieve a 1 Percent Reduction in BOD$_5$ Discharge, 1972 Conditions, Five-Commune Area.[a]

Physical Measure	Activity Category	Present value of costs, 10^6 N.D.,[b] per 1% reduction in BOD$_5$ discharge	Cost Effective Ranking	
			With respect to single measure	With respect to all measure-activity combinations
Primary sedimentation/activated sludge, collective wastewater treatment plant	All activities presently connected to sewer system	3.1	1	1
	All activities in connection area I	29.3	2	3
	All activities in connection area II	99.7	3	4
	All activities in connection area III	1,017	5	6
	All activities in connection area IV	659	4	5
	All activities in connection area V	1,315	6	7
On-site primary sedimentation/aerated stabilization waste-water treatment plant	Twelve outlying industrial plants			
	Primary sedimentation at all 12 plants plus aerated stabilization at 6 of the 12 plants	25.9	--	2

a All values rounded.
b Costs are net and in 1974 N.D.; 40 year period of analysis; 11% discount rate.

achieve a 1 percent reduction in ambient SO_2 concentration for each of the physical measure-fuel usage activity combinations shown in table 29. Table 32 shows the present value of costs to achieve a 1 percent reduction in BOD_5 discharges for the physical measure-activity combinations shown in tables 23 and 24 of chapter VI.

Combinations of Physical Measures and Their Effects

The cost effectiveness, as defined above, of each physical measure-activity combination suggests the sequence of addition of physical measures as components of the minimum cost set of physical measures to meet the specified E.Q. targets. Physical measure-activity combinations are added successively, beginning with the most cost effective physical measure-activity combination and continuing until all target values are met or exceeded. Each physical measure-activity category is added in its entirety. That is, the physical measure is applied to all individual activities in the category and the costs for such application are computed. If more analytical resources had been available, it would have been possible to subdivide activity categories and apply different physical measures to the subdivisions.

The cost effectiveness index used related to only one residual or one E.Q. indicator and, therefore, ignored the joint effects of a physical measure on more than one residual. Thus, the combinations of physical measures analyzed should not be limited to those indicated by the cost effectiveness index. Application of the index, however, resulted in a more efficient use of analytical resources than a complete enumeration and analysis of all possible combinations. The analytical procedure used--the search method--is illustrated in figure 22.

Results of the Analysis

The physical measure-activity combinations comprising the least cost combination of physical measures, denoted as combination I, are listed in table 33, along with their effects and associated costs. The E.Q. targets, the E.Q. levels actually achieved, and the present value of costs for combination I are shown in table 34. By limiting the application of physical measures to activity subcategories as a whole, and because some physical measures simultaneously affect more than one residual, some target values cannot be met exactly. The discrete physical measures of combination I: (1) do not enable achieving the HC, CO and NO_x targets; (2) result in considerably exceeding the TSP target; (3) result in moderately exceeding the SO_2 target; (4) result in not quite meeting the BOD_5 target and meeting the TSS target; and (5) result in disposal of all MSR in a good quality sanitary land-fill. Combination I is estimated to cost about 3.4 billion 1974 N.D.

203

Figure 22. Flow Diagram of Search Method as Applied in Ljubljana Study

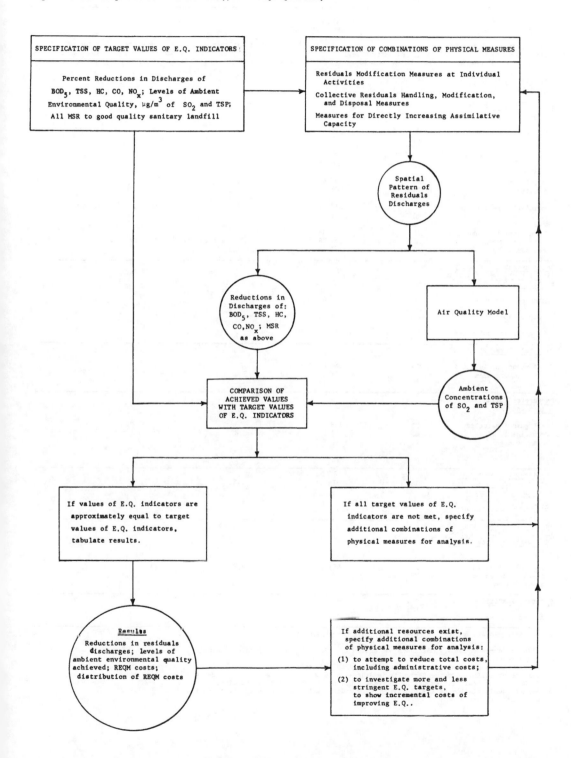

Table 33. Physical Measure Activity Combinations Comprising the Least Cost Combination of Physical Measures to Achieve E.Q. Targets, 1972 Conditions, Five-Commune Area[a]

Physical Measure-Activity Combination	E.Q. Indicator	Mean degree of discharge reduction, or improvement in AEQ in region	Present value of capital costs, 10^6 N.D.	Present value of O&M costs, 10^6 N.D.	Total present value of costs, 10^6 N.D.
Substitution of light liquid fuel with 1% sulfur, essentially no ash, for coal with 2% sulfur, 12.5% ash					
Chemical industry	SO_2 TSP	7% AI 8% AI	4	18	22
Pulp & paper industry	SO_2 TSP	19% AI 30% AI	5	95	100
Metal processing industry	SO_2 TSP	6% AI 10% AI	3	28	31
Multi-flat residences	SO_2 TSP	11% AI 4% AI	6	100	106
Total effect of above on other gaseous residuals	CO HC NO_x	5% DR 15% DR 5% DR	No costs in addition to above costs		
Desulfurization of heavy liquid fuel from 3% sulfur to 0.5% sulfur					
Chemical industry	SO_2	7% AI	b	25	25
Pulp & paper industry	SO_2	8% AI	b	58	58
Single-flat residences	SO_2	1% AI	b	13	13
Park/Ride Level II					
13% reduction in Vkt in five-commune area	CO HC NO_x	16% DR 9% DR 1% DR	300	480	780
Carburetor idling adjustment					
All gasoline-fueled vehicles in five-commune area	CO HC NO_x	17% DR 1% DR -- DR[c]	2	106	108
On-site primary sedimentation/aerated stabilization wastewater treatment plant for 12 outlying industrial plants[d]	BOD_5 TSS	23% DR 7% DR	43	18	61
Primary sedimentation/activated sludge, collective wastewater treatment plant					
All activities in areas feasible to connect to sewer system	BOD_5 TSS	53% DR 73% DR	1732	373	2105
Sanitary landfill All activities	All MSR	--	15	11	26

Abbreviations: AI, ambient improvement; DR, discharge reduction

a All values are rounded. Costs are net and in 1974 N.D.; 40-year period of analysis; 11% discount rate.
b Capital costs of desulfurization are reflected in O&M costs.
c Carburetor idling adjustment results in decreases in CO and HC discharges, increase in NO discharges. The amount of increase could not be estimated at the time.
d Primary sedimentation at six plants; primary sedimentation plus aerated stabilization basin at six plants.

Table 34. Environmental Quality Targets, Estimated Environmental
Quality Levels Achieved, and Estimated Costs for Least
Cost Combination of Physical Measures, 1972 Conditions,
Five-Commune Area[a]

E.Q. Indicator	E.Q. Target	Mean E.Q. level achieved	Present value of costs, 10^6 N.D.
SO_2	150 $\mu g/m^3$	135 $\mu g/m^3$	} 360
TSP	150 $\mu g/m^3$	74 $\mu g/m^3$	
BOD_5	80% reduction	76% reduction	} 2170
TSS	80% reduction	80% reduction	
CO	50% reduction	35% reduction	
HC	50% reduction	25% reduction	} 890
NO_x	10% reduction	6% reduction	
MSR	All MSR into good quality sanitary landfill	All MSR into good quality sanitary landfill	30
	Present value of total cost		3450

a All values are rounded. Costs are net and in 1974 N.D.;
40-year period of analysis; 11% discount rate.

Because the last increments of physical measures added to reduce TSS, BOD_5, and SO_2 discharges are relatively expensive per unit of discharge reduction achieved, additional combinations of physical measures were developed to illustrate the costs of achieving different levels of environmental quality. In addition to combination I, four other combinations of physical measures were developed. In all combinations all MSR generated are deposited in good quality sanitary landfill and the same degrees of discharge reduction are achieved for CO, HC, and NO_x. The E.Q. levels achieved by, and the costs of these combinations are shown in Table 35, along with the corresponding results for combination I. For combination II, the reductions in TSS and BOD_5 discharges achieved are slightly less than for combination I, a result accomplished by eliminating the last two increments of measures shown in figure 23. (In figure 23 increments are added in relation to the least cost physical measures for reducing BOD_5 discharges. The simultaneous reductions in TSS discharges are also shown.) The ambient concentrations of SO_2 and TSP achieved are the same as for combination I. Costs of combination II are estimated to be about 2.4 billion 1974 N.D., or about 70 percent of the costs for combination I, with only slightly lower reductions in TSS and BOD_5 discharges.

For combinations III and IV, the ambient concentrations of SO_2 and TSP achieved are higher than, and the reductions in TSS and BOD_5 discharges achieved are less than, those achieved by either combination I or combination II. In combination III the last three increments of physical measures to reduce discharges of BOD_5 and TSS, and the last increment of combination I to reduce SO_2 discharges--physical measure-

Table 35. Environmental Quality Levels Achieved and Corresponding Present Values of Costs for Various Combinations of Physical Measures, 1972 Conditions, Five-Commune Area[a]

E.Q. Indicator	Combination I		Combination II		Combination III		Combination IV		Combination V	
	Mean E.Q. level achieved	P.V. of costs, 10^6 N.D.	Mean E.Q. level achieved	P.V. of costs, 10^6 N.D.	Mean E.Q. level achieved	P.V. of costs, 10^6 N.D.	Mean E.Q. level achieved	P.V. of costs, 10^6 N.D.	Mean E.Q. level achieved	P.V. of costs, 10^6 N.D.
SO_2 TSP	135 µg/m³ 74 µg/m³	360	135 µg/m³ 74 µg/m³	360	170 µg/m³ 80 µg/m³	250	200 µg/m³ 80 µg/m³	180	110 µg/m³	470
CO HC NO_x	35% DR 25% DR 6% DR	890	35% DR 25% DR 6% DR	890	35% DR 25% DR 6% DR	890	35% DR 25% DR 6% DR	890	35% DR 25% DR 6% DR	890
TSS BOD_5	80% DR 76% DR	2170	76% DR 71% DR	1180	74% DR 69% DR	840	63% DR 61% DR	420	76% DR 71% DR	1180
MSR	All MSR into good quality sanitary landfill	30	All MSR into good quality sanitary landfill	30	All MSR into good quality sanitary landfill	30	All MSR into good quality sanitary landfill	30	All MSR into good quality sanitary landfill	30
Total Costs		3450		2460		2010		1520		2570

Abbreviation: DR, discharge reduction

[a] All values are rounded. Costs are net and in 1974 N.D.; 40-year period of analysis; 11% discount rate.

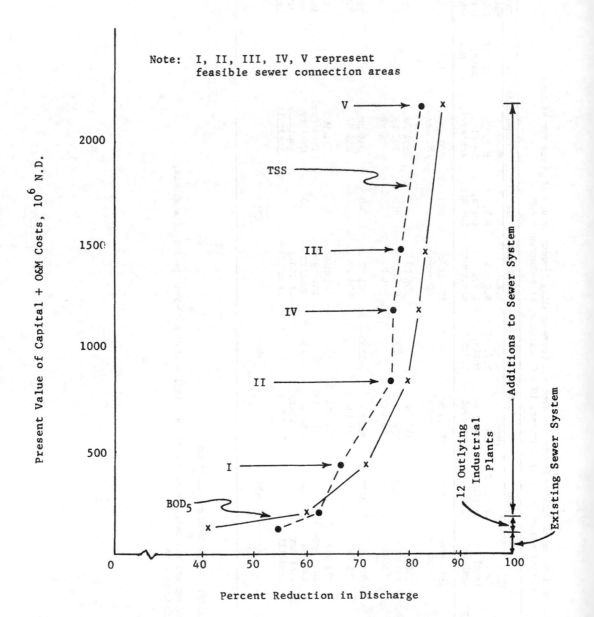

Figure 23.   Discharge Reduction-Cost Functions for BOD$_5$ and TSS, 1972 Conditions, Five-Commune Area[a]

a   Costs are net and 1974 N.D.;   40-year time period of analysis; 11% discount rate.

activity combination 7A in figure 24--are eliminated. In combination IV

the last four increments of measures to reduce BOD_5 and TSS discharges

and the last three increments of combination I to reduce SO_2 discharges--

7A, 6B, and 3B in figure 24--are eliminated. For combination IV, the

ambient concentration of SO_2 achieved is higher than, and the reductions

in TSS and BOD_5 discharges achieved are less than, the corresponding

levels for combination III; the ambient concentration of TSP is the same

for both combination III and combination IV. These decreases in environ-

mental quality reflect decreases in costs, the estimated costs for combi-

nation III being about 2.0 billion 1974 N.D., about 60 percent of the

estimated costs for combination I, and the estimated costs for combina-

tion IV being about 1.5 billion 1974 N.D., about 45 percent of the costs

for combination I.

Combination V was developed because of the concern in the five-

commune area for improving ambient air quality. For combination V, the

ambient concentrations of SO_2 and TSP achieved are lower, and the reduc-

tions in discharges of TSS and BOD_5 are slightly less, than for combina-

tion I. The estimated costs for combination V are about 2.6 billion 1974

N.D., about 75 percent of those for combination I.

It should be emphasized that all of the costs shown in the tables

exclude administrative costs. The implementation of any combination of

physical measures involves administrative activities such as monitoring,

inspection, analysis of samples, collecting charges and fines. Such

activities involve costs, which depend both on the components of the

combination and the implementation incentives used to induce the adoption

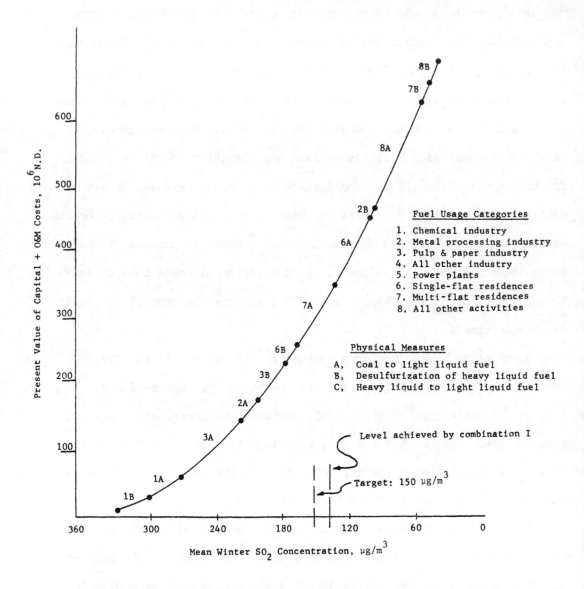

Figure 24. Relationship between Mean Winter Concentration of SO_2 and Discharge Reduction Costs, 1972 Conditions, Five-Commune Area[a]

a   Costs are net and in 1974 N.D.; 40 year time period of analysis; 11% discount rate.

of the various physical measures involved. These administrative costs comprise an integral part of REQM costs, and can be substantial.

Implications of Results

What do the results of analyzing combinations of physical measures imply for REQM and for the allocation of resources to REQM in the five-commune area? Because each combination must be integrated with implementation incentives and institutional arrangements to form an REQM strategy, as discussed in the next chapter, only a partial answer to the question is possible at this point. Further, because monetary damage/benefit functions do not exist, that combination of physical measures which maximizes net benefits cannot be determined. Nevertheless, the results clearly have implications with respect to REQM in the five-commune area. The order in which they are discussed does not imply the order of importance.

1. Each increment of reduction in TSS and BOD_5 discharges above the 70 to 80 percent range, by means of end-of-pipe measures, e.g., standard physical-biological sewage treatment, results in rapidly increasing incremental costs. Even if the benefits from greater than 70 to 80 percent reduction in BOD_5 and TSS discharges were high, efficient utilization of scarce resources suggests that REQM efforts should be oriented toward physical measures to reduce BOD_5 and TSS generation, e.g., by in-plant and/or in-household changes, rather than toward end-of-pipe modification. This orientation is just as relevant for most other residuals and will become even more important as the economy of the area grows, and more demands are made on finite environmental resources.

2. The relatively low cost of good quality sanitary landfilling suggests that this physical measure should be a part of any REQM strategy for the area. The benefits in terms of preventing adverse impacts on ground water quality, maintaining a positive visual appearance in the area, and protecting public health are likely to be significantly greater than the costs.

3. The two physical measures for reducing CO, HC, and NO_x discharges from vehicular movements achieve only a relatively small decrease in such discharges. This suggests that major changes in the transport system within the five-commune area will be necessary if any significant improvement in these measures of environmental quality is to be achieved. Major changes in motive power of individual vehicles and a major shift to mass transportation, with changed motive power, will be necessary.[46]

4. Shifting to off-site supply for central heating can have negative impacts on ambient SO_2 concentrations, if more than two of the feasible connection areas are connected. Three factors account for this. First, the shift to off-site supply results in an increase in the total amount of energy conversion required for heating purposes. Second, low quality solid fuel is, and would be, burned in the power plants which would provide most of the energy for off-site central heating, in contrast to high quality liquid fuel and brown coal which are presently the

---

[46]Although the Park/Ride measure for reducing discharges of CO, HC, and NO_x has relatively small impacts on such discharges, other non-REQM related benefits result, namely, reduced congestion in the central city and savings in energy. It may be that these benefits alone might justify the Park/Ride measure.

principal fuels used in on-site heating installations. The quantity of
fuel used by power plants is about half the total fuel combusted in all
activities, other than transportation, in the five-commune area. To
shift to light liquid fuels or to desulfurized heavy liquid fuel would
involve major expenditures at power plants, given the present sources
and qualities of fuel and large existing investments in coal combustion
technology. In addition, only low energy content lignite coal and brown
coal are in good supply. Liquid fuels, most of which are imported, are
not in ample supply to satisfy the greatly increased demand which would
occur if the off-site supply for central heating alternative were
adopted. Third, on-site wet scrubbing and limestone injection to reduce SO_2 discharges from power plants have not been considered as feasible
discharge reduction alternatives because of excessive costs and technical
difficulties in obtaining the appropriate equipment in Yugoslavia.
Shifting to off-site central heating does reduce particulate discharges,
but it is not the least cost measure for reducing such discharges.

5. The winter season averaging period used in the analysis of
ambient air quality masks the effects of short periods of poor air
quality of several days duration, which occur quite frequently in the
five-commune area. First, even though the monthly mean ambient concentrations of SO_2 and TSP meet their respective standards, concentrations
substantially higher than the standards can occur for short periods.
Second, analysis of the historical data showed that in some months the
mean monthly concentrations of SO_2 and TSP were 25 percent higher than

the respective standards because of a few days of high concentrations, rather than because of concentrations continuously above the standard throughout the month.

The physical measures investigated will effectively reduce ambient concentrations on the average. The minimum cost combination to achieve the ambient SO_2 standard required, for example: (1) plants in the chemical industry to substitute light liquid fuel for the coal presently burned and substitute desulfurized heavy liquid fuel in place of the high sulfur heavy liquid fuels now burned; and (2) some of the plants in the pulp and paper industry to change from coal to light liquid fuel. These measures, however, will not enable achieving the ambient SO_2 standard during short-period, critical conditions. The only physical measure which can be applied intermittently is the substitution of desulfurized heavy liquid fuel for the high sulfur heavy liquid fuels. If such action is insufficient to enable achieving ambient air quality standards, the only recourse is to reduce the levels of activity during critical periods.

Chapter IX

REQM STRATEGIES AND CONCLUDING OBSERVATIONS

The discussion in the preceding chapter described how the first component of REQM strategies--physical measures--was analyzed in relation to
alternative levels of environmental quality in the five-commune area. The
first part of this chapter discusses the relationship between those
physical measures and the other two components of REQM strategies, institutional arrangements and implementation incentives, in the Yugoslav
context. In the second part of the chapter are presented some concluding
observations on the utility of the study for REQM in the five-commune
area and on some necessary next steps for REQM analysis in the area.

REQM Strategies for the Five-Commune Area

The separation of the discussion of physical measures from the discussion of the other two elements of REQM strategies does not mean that
the physical measures analyzed were chosen without consideration of implementation incentives and institutional arrangements. In accord with
the goal of making the analysis as Yugoslav-oriented as possible, consideration was given to the institutions and incentives available for
inducing adoption of physical measures in selecting the alternative
physical measures to be analyzed. This meant identifying organizations
and individuals who could provide the necessary information and assist
in selecting and developing physical measures for analysis.

Selecting Physical Measures

Selecting physical measures for improving ambient air quality, particularly with respect to SO_2 and TSP, received the most scrutiny for three reasons. One, the air quality problem is perceived by many people to be the most critical environmental quality problem[47] in the five-commune area. Two, most government interest, as reflected principally by the Slovene Republic Committee for Air Quality, has been in improving air quality. Three, it was possible to develop an air quality model as described in chapter VII, which enabled analyzing the quantitative impacts of alternative physical measures on ambient air quality.

The Committee for Air Quality assisted in identifying physical measures for reducing discharges of SO_2 and particulates. Two sessions were held with the Committee. In the first, previous analyses by the project staff with the air quality model and data from published material were presented to the Committee and other interested parties.[48] The second was a closed working session to delineate physical measures which were

---

[47]For documentation, see: D.E. Kromm, "Responses to Air Pollution in Ljubljana, Yugoslavia," Annuals of the Association of American Geographers, vol. 63, no. 2, June 1973.

[48]These included Snaga, Kanalizacija, Zavod za Vodno Gospodarstvo SRS (Regional Water Planning Authority of the Slovene Republic), Smelt (private engineering consulting firm), Hidrometeoroloski Zavod SRD (the Slovene Republic Office for Meteorology), Biotehniska Fakutteta (Biology Department, University of Ljubljana), Oddelek za Meteorologijo (School of Meteorology, University of Ljubljana).

considered feasible to implement. In this working session[49] a question-

naire containing a list of potential physical measures was presented to

the Committee. After much discussion, three measures were selected for

analysis: fuel substitutions, off-site central heating from power plants,

and the use of stack-gas cleaning equipment. Fuel desulfurization, a

physical measure which was not on the list presented to the Committee,

was later selected for analysis by the project staff. The specific de-

tails of each measure were worked out by the project staff.

Physical measures to reduce discharges of CO, HC, and NO_x from ve-

hicular movements were suggested by an authority[50] at the University of

Ljubljana. The basic approach adopted in Yugoslavia for reducing dis-

charges from vehicular movements is consistent with the general European

approach, namely, to await changes in engine design rather than add emis-

sion control devices to existing engine designs as in the United States.

Consequently, only two physical measures were identified: (1) regular

adjustment of carburetor idling; and (2) a Park/Ride bus transit system

between the periphery and the center of the city. Both of these physical

measures were considered operationally feasible by both the Ljubljana

---

[49]The following members of the Committee attended this session held
on June 6, 1974: Peter Novak, dip. ing. stroj.: Slavko Verhovnik, dip.
fizik; Dr. Bogdan Sicherl; Anton Eubanc, ing.; Janez Novak, dip. ing.;
Lundor Adamic, ing.; Borjan Paradix, dip. ing. meteor.; Dr. Zdravko
Petkovsek, dip. meteor.; and Prof. Andrej Jocevar. Miael Gruden, dip.
ing., not a member of the Committee, also attended.

[50]Dr. J. Pavletiv, Faculteta za Strojnistvo, University of Ljubljana.

Urban Planning Authority (LUZ) and Yugoslav staff members of the project.

Neither of these physical measures requires a basic change in technology.

The only physical measures considered feasible to reduce discharges

of BOD_5 and TSS were standard wastewater treatment plants, both at twelve

specific industrial operations and at a regional plant to handle most of

the remaining load generated in the five-commune area. Even if analytical

resources had been available to investigate physical measures within

activities to reduce BOD_5 and TSS discharges, such as by changes in pro-

duction processes and increased materials recovery, such measures were

considered not feasible to implement at the time[51] or in the near future.

With respect to physical measures for handling MSR, relevant municipal

and private agencies were contacted to determine which of the measures

currently available in the U.S. and Western Europe could be considered

feasible for the five-commune area. Because of constraints on obtaining

certain types of technological options, import regulations, and available

cost information, only sanitary land filling of MSR was considered feasible

at this time or in the near future.

Institutional Arrangements

REQM involves the set of activities described in chapter II, which

activities must be performed in order to produce and maintain the

---

[51]This was indicated in discussions with Kanalizacija, Zavod za
Vodno Gospodarstvo SRS, Mestni Vodovod, LUZ, and Splosna Vodna Skupnosti
(Regional Water Self-Management Community).

desired product of improved AEQ. In any society those activities are likely to be performed by a set of institutions, agencies, enterprises, operating at all levels of government. No <u>single</u> agency will perform all activities. The particular institutional arrangement for REQM will inevitably and necessarily reflect the socio-political, economic, and cultural systems of the society. Hence it would be irrelevant to pre-scribe some "normative" concept of institutional organization for REQM in Yugoslavia--including the specification of the loci of authority for imposing implementation incentives--without very extensive analysis of the institutional/governmental structure of the country.

In any REQM context, inducing the residuals generating activities to apply physical measures depends on the existence of an institutional structure which has authority to impose implementation incentives which will induce the desired actions. Each physical measure identified for each activity must be related to one or more implementation incentives and, in turn, to one or more institutions.

Before identifying specific physical measure-institution-implementa-tion incentive combinations for the five-commune area, five "facts of life" concerning Yugoslav governmental structure should be stressed.

First, no agency specifically responsible for REQM appears to exist in Yugoslavia at any level of government, although various agencies and enterprises at the commune, multi-commune, republic, and national levels carry out parts of REQM. Second, the ultimate responsibility for achiev-ing and maintaining environmental quality is delegated to the communes.

The primary role at present of the republic and federal governments is the establishment of "guideline" laws. Third, workers' management organizations at the commune and multi-commune levels must initiate, develop, and enforce REQM strategies under federal and republic "guideline" laws. In Yugoslavia, each enterprise in the social sector is operated in a workers' management framework which has some form of self-government and employs profit sharing as an incentive mechanism.

Fourth, the commune is the basic unit of government in Yugoslavia and hence is the logical governmental "building block" for REQM. Communes, as the basic political units, send delegates to the three other levels of government: a city assembly, e.g., the Ljubljana assembly; the republic assembly, e.g., Slovenia; and the national assembly. Communes collect virtually all tax revenues--except those from taxes on foreign trade--and then make contributions to the city and other levels of government. These taxes are on personal and enterprise income, and account for most of the revenue obtained. The city assembly, comprised of delegates from each of the five communes, then allocates the available resources to provide various public services. For example, the city contracts with enterprises such as Kanalizacija and Snaga to carry out sewage collection and disposal and solid residuals collection and disposal, respectively. Because communes are small in size, it is relatively easy to identify combinations of communes which would yield political boundaries corresponding more or less to boundaries of REQM problems.[52] The five-commune area coincides reasonably well with the gaseous and solid residuals management problem sheds. But as noted in chapter VII, for efficient management of liquid residuals, the five-commune area is inadequate; the logical water quality management unit is the set of communes which comprise the Upper Sava River Basin.

Fifth, the governmental structure in Yugoslavia changed significantly with the constitutional revisions of 1946, 1952, 1963, and 1974. Such evolution is to be expected in a rapidly developing and an evolving society. But it means that delineating institution-implementation incentive combinations is likely to be provisional.

Regardless of the changing relationships in the structure of Yugoslav institutions, it appears that the critical link in implementation will be to induce the communes to action, both to impose implementation incentives on individual activities and to undertake various collective measures for residuals handling and disposal. Because the communes must provide their own resources to affect AEQ, there may be little incentive to action unless serious ambient environmental quality conditions are perceived, as may be the case where receptors as a whole in the five-commune area perceive significant damages to human health and property from adverse air quality.

---

[52]If improved AEQ is considered a public service, establishment of a self-managing community of interest for REQM would appear to be logical under the 1974 Constitution. See Kasoff, M.J., "Local Government in Yugoslavia and Constitutional Reform of 1974, a Case Study of Ljubljana," American Institute of Planners Journal vol. 42, no. 4, October 1976, pp. 339-409.

There may be little incentive to action because the costs to improve AEQ
may be perceived as being too high and/or the resources available to the
communes to deal with an REQM problem may be insufficient or nonexistent.
Given the decentralized structure of Yugoslav society, there are likely
to be few funds and/or few mechanisms to channel funds from higher levels
of government to the commune level, as is frequently done in more cen-
tralized systems. At present, neither the federal government nor the
republic government has significant "carrots or sticks" to induce communes
to act. Clearly, unless it is perceived in a society that there is an
environmental quality problem, there is no stimulus for institutional
development and subsequent action.

Implementation Incentives

Implementation incentives, as defined in chapter II, are the means
by which residuals generating and discharging activities and collective
residuals handling and modification activities are induced to adopt
physical measures for reducing discharges. Five types of implementation
incentives to induce an activity to take a desired action were identified
and defined in chapter II: regulatory, economic, administrative, judicial,
and educational/informational. All implementation incentives require
some institutional arrangement which provides authority to one or more
agencies to impose the incentives on the relevant activities. Because of
the cross-cultural context and the evolutionary nature of Yugoslav institutions, explicit analysis of implementation incentives was not

included in the study. However, it is useful to identify possible implementation incentives-physical measure-institutional arrangement combinations related to the specific physical measures and activities delineated
in previous chapters.

Fuel substitution to reduce SO_2 and particulate discharges could
presumably be implemented by a regulation promulgated by the commune and/or
republic government(s), e.g., a requirement that particular activities
use fuels with maximum limits on sulfur and ash contents. A related incentive would be for the republic government to subsidize the use of desulfurized fuel. Similarly, the installation of cyclonic filters could
be required, by regulation by the commune, on all boilers exceeding a
specific capacity and using solid fuel. Alternatively, an effluent charge
could be levied by the communes on each kilogram of sulfur and each kilogram of particulates discharged by each activity. The charge would then
be a cost of production, similar to the costs of other factor inputs,
such as raw materials and labor. An enterprise seeking to minimize production costs would determine the least expensive way to minimize such
costs, e.g., by changing fuel, installing cyclonic filters, changing
production process. If the charge were sufficiently high, it would induce the adoption of one or more of the physical measures feasible for
the particular activity. In addition, the commune governments could provide loans to or make credit arrangements for enterprises to help finance
the installation of cyclonic filters or other physical measures to reduce
SO_2 and particulate discharges.

To reduce discharges of HC and CO from vehicular movements, implementation of the carburetor idling adjustment would require a system for monitoring and maintaining the performance of all gasoline-fueled vehicles in the five-commune area. Such monitoring and maintenance could be combined with regular safety monitoring and maintenance in communal inspection stations. Implementation of the Park/Ride measure to reduce discharges of HC, CO, and NO_x requires inducing individuals to forego the use of their vehicles in the central city. This could be accomplished by adoption by the city assembly of Ljubljana of some combination, or all, of the following: (1) decrease the number of available parking spaces in the central city; (2) install meters on all parking spaces in the central city; (3) increase the metered parking rates; and (4) increase the frequency of monitoring the meters, accompanied by enforcement of fines for illegal parking.

Construction and operation of wastewater treatment plants at the twelve outlying industrial activities to reduce TSS and BOD_5 discharges could be induced by a regulation by the commune requiring the construction of such plants by each enterprise, accompanied by a system of random inspections to check that the plants were being properly operated after construction and a system of fines for failure to meet the discharge standards. Alternatively, as for SO_2 and particulates, an effluent charge could be levied on the discharge of each kilogram of TSS and BOD_5. If the charges were sufficiently high, enterprises would be induced to adopt some physical measures for reducing their discharges. Commune governments could provide loans to or make credit arrangements for enterprises to help finance the installation of physical measures. The regional wastewater treatment plant could be implemented by the communes in the same manner as with other infrastructure facilities.

Concluding Observations on the Analysis
for REQM in the Five-Commune Area

Utility of the Study

The results of the study have immediate utility for REQM decision making in the five-commune area. First, the information developed clarifies and resolves issues concerning: (a) the effect on ambient air quality of off-site central heating; (b) the relative contributions of industrial activities and residences to air pollution as measured by ambient SO_2 and TSP concentrations; (c) the effects and costs of connecting activities in the five-commune area to the municipal sewage treatment plant during the next two decades; and (d) the effects of reducing vehicular movements on discharges of HC, CO, and NO_x in the five-commune area.

Off-site central heating is not the key to reducing ambient SO_2 and TSP concentrations, as was widely hypothesized. If all feasible connection areas were connected to off-site central heating facilities, SO_2 discharges in the region would actually increase, and particulate discharges would only decrease slightly and at very substantial costs.

Industrial activities contribute more than residential activities to ambient concentrations of SO_2 and TSP, contrary to widespread opinion that the opposite is the case.

Because substantial differences were found in the unit costs of reducing BOD_5 and TSS discharges among the areas proposed for connection to the single proposed municipal treatment plant, efficient use of available resources would be achieved by using the criterion, amount of discharge reduction achieved per 10^6 N.D. expended, to determine the sequence of addition of connection areas to the sewer system.

The two physical measures analyzed to reduce discharges of CO and HC from vehicular movements achieve only small decreases in such discharges relative to what had been anticipated. This result suggests that: (a) major changes in the transportation system within the five-commune area will be required if significant reductions in CO and HC discharges are to be achieved; and (b) the adoption of physical measures directed at energy conversion in activities other than vehicular movement will be necessary if a significant reduction in NO_x discharges is to be achieved, because NO_x discharges from vehicular movements account for less than ten percent of NO_x discharges in the five-commune area.

These conclusions appear valid despite the fact that the analysis on which they are based used 1972 data. Using 1972 data as the base conditions for the analysis means that the absolute values of residuals generated and discharged, REQM costs, and ambient concentrations are probably relevant not much beyond 1975-1976. However, updating the data would not likely shift: (a) the relative ranking of the physical measures in terms of cost-effectiveness; and (b) the stated conclusions.

Second, the study has provided a framework and one methodology for continuing and improving analysis for REQM in the five-commune area. The municipal MSR collection agency could use data generated in the study to improve routing schedules for collecting MSR and to investigate incineration or energy recovery technologies for possible future adoption. A method for estimating costs of connecting sewer connection areas to the system and a procedure for deciding on the sequence for adding areas were developed with the municipal sewer agency. These can be used for evaluating alternative regional strategies for managing liquid residuals in the area. A similar system for determining the cost-effectiveness of incremental additions to the energy distribution system was developed with the central power plant agency.

Third, even though a limited number of physical measures to reduce discharges of residuals were analyzed, especially with respect to industrial activities, the results show what degrees of reduction in the discharges of four major residuals--SO_2, particulates, BOD_5, and TSS--can be achieved for different expenditures. Such information, in addition to the perceived relative importance of those residuals, provides a basis for selecting an REQM strategy.

Fourth, a common view in the five-commune area is that AEQ can only be improved and maintained at the expense of economic development. Of course in no society are resources unlimited. The demand for improved AEQ is but one of the demands for goods and services desired by a society. Thus, decisions must be made about the allocation of resources among the various sectors. The assumption is often made--in all societies--that improving AEQ will require large expenditures. Depending on the REQM strategy adopted, this often is not the case, as detailed analyses of residuals management in industry have shown.[52] This point is made because of the view expressed in meetings with the Slovene Republic Committee on Air Quality, and in meetings with other groups, that reducing the discharges of residuals from industrial operations can only result in decreased productivity at increased costs. In fact, having to re-analyze the combination of factor inputs to produce a product or service--as a result of having to reduce residuals discharges--has led in many cases to decreased production costs, e.g., more efficient overall production.

Because this is an important issue, the annual cost of each physical measure applied to each relevant industrial subcategory was calculated. For each industrial subcategory the percentage these annual costs were of the wholesale market value of product outputs for the subcategory in the base year, 1972, was computed. The largest percentage found was 1.3 percent for substituting light liquid fuel for coal in the pulp and paper industry. All other percentages for industrial subcategories were less than 1 percent, indicating that industrial activities should be capable of paying the REQM costs with little impact on their

---

[52]See B.T. Bower, "Studies of Residuals Management in Industry," in E.S. Mills, ed., Economic Analysis of Environmental Problems (New York: National Bureau of Economic Research, 1975), pp. 275-320.

competitive positions. This is particularly true because the costs for industrial activities were based solely on end-of-pipe and fuel substitution measures. Very often there are in-plant changes which achieve the same reduction in residuals discharges with substantially lower costs.

What Next in Analysis for REQM in the Five-Commune Area?

The work reported herein represents a base upon which to build REQM analysis for the five-commune area. A number of additional activities are required to broaden and deepen the analysis in order to make it more useful for REQM decisions.

1. Specific implementation incentives and the necessary associated institutional arrangement(s) to impose them must be identified for each of the physical measure-activity subcategory combinations delineated in the study. When identifying implementation incentives which might be imposed, it is particularly important to be certain that an implementation incentive applied at one level of government, e.g., commune, is consistent with, and not counterproductive to, an implementation incentive applied at another level of government, e.g., republic or national. Similarly, physical measure/implementation incentives must be consistent with other national policies. For example, suppose implementation incentives, such as loans to and credit arrangements for enterprises, were adopted at the commune level to induce various activities using solid fuels in their boilers to change to liquid fueled boilers to improve air quality. Such an implementation incentive at the commune level would be

inconsistent with incentives at the national level directed at increasing the use of domestic coal reserves and reducing imports of foreign liquid fuels, e.g., a subsidy for domestic coal and a tariff on imports of foreign oil. An REQM strategy which posited the use of equipment available only outside the country might be inconsistent with import policy.

2. Not only must implementation incentives be identified for each of the physical measure-activity subcategory combinations, but the effectiveness of the incentives over time must be monitored. That is, how well in fact do they induce action by the activity. The responses to the implementation incentives also need to be monitored in order to determine their impacts on technological innovation.

3. The range of physical measures for reducing discharges should be expanded, in order to reduce as much as possible the costs of improving AEQ. The physical measures considered for reducing discharges from activities other than vehicular movements were limited to end-of-pipe facilities and fuel substitution. No alternatives such as changing production technology, increasing materials and/or energy recovery, by-product production, changing product mix, changing product specification, were considered, although they are often less expensive than the end-of-pipe and fuel substitution measures.

Expanding the range of physical measures analyzed is particularly important where the incremental costs of reducing discharges are high. For example, the incremental costs of connecting each of the last three sewer connection areas to the proposed municipal wastewater treatment plant are very high. Alternatives involving combinations such as a smaller central plant with one or two satellite plants should be analyzed. Although some economies of scale would be lost, sewage transport costs would be reduced, as would the magnitude of the final load imposed at one place on the river.

4. The range of activities analyzed should be expanded. Ambient water quality in the area is affected by residuals discharged in runoff from urban areas and agricultural lands. How important these sources are can only be determined by explicit analysis.

5. The number of residuals analyzed should be expanded. The present analysis was limited by the available data and resources. But other residuals, such as toxics, may be important in the area. Effort needs to be expended in order to determine what other residuals, if any, are important.

6. The time horizon of the analysis should be extended to the end of the century. This would enable looking explicitly at the environmental quality implications of alternative spatial patterns of activities, alternative levels of activities, alternative types of production processes. The impacts on REQM costs to achieve and maintain desired levels of AEQ, and the distribution of those costs, could be estimated. Defining quantitatively the interrelationship among land use pattern, transportation system, and environmental quality would be of help in decision making.

7. Each of the physical measure-implementation incentive-activity subcategory combinations should be evaluated on the basis of the criteria delineated in chapter II, and the relative weights attached to those criteria by the relevant Yugoslav authorities. Such evaluation would provide a more adequate basis for selecting REQM strategies than the single criterion of cost-effectiveness used in this report.

All of the foregoing depend on the existence of a continuous analysis and planning process and an associated data collection program. This is particularly important in a rapidly developing economy where factor input prices, tax policies, social tastes, foreign exchange constraints, production technology, raw materials, institutional arrangements, may change frequently. Thus, some institutional mechanism must be established for the five-commune area to perform the continuous REQM analysis.

Hopefully, the present study of REQM in the five-commune area has demonstrated the utility of the REQM framework and of the systematic analysis of REQM problems, provided specific data for use in decision making, and established a base for an ongoing, continuous activity of REQM analysis.